STONE

Stone Building Materials, Construction
and Associated Component Systems:
Their Decay and Treatment

ENGLISH HERITAGE

ENGLISH HERITAGE RESEARCH TRANSACTIONS
RESEARCH AND CASE STUDIES IN ARCHITECTURAL CONSERVATION

STONE

Stone Building Materials, Construction
and Associated Component Systems:
Their Decay and Treatment

EDITED BY
John Fidler

Volume **2**

January 2002

© Copyright (text) 2002 English Heritage
© Copyright (illustrations) 2002 the authors or other copyright holders as stated in the captions

All rights reserved. No part of this publication may be reproduced, stored in a retrieval system or transmitted in any form or by any means, electronic, mechanical photocopying, recording or otherwise, without the prior written permission of the copyright owner and the publisher.

First published by James & James (Science Publishers) Ltd, 35–37 William Road, London NW1 3ER, UK

A catalogue record for this book is available from the British Library
ISBN 1-873936-63-X
ISSN 1461 8613

Volume editor: John Fidler RIBA, English Heritage
Series editor: David Mason, English Heritage
Consultant editor: Kate Macdonald

Printed in the UK by Hobbs The Printers

Disclaimer
Unless otherwise stated, the conservation treatments and repair methodologies reported in this volume are not intended as specifications for remedial work. English Heritage, its agents and publisher cannot be held responsible for any misuse or misapplication of information contained in this publication.
The inclusion of the name of any company, group or individual, or of any product or service in this publication should not be regarded as either a recommendation or endorsement by English Heritage or its agents.

Accuracy of information
While every effort has been made to ensure faithful reproduction of the original or amended text from authors in this volume, English Heritage and the publisher accept no responsibility for the accuracy of the data produced in or omitted from this publication.

Front cover:
Jane Scarrow, architectural conservator, cleaning masonry on the Wellington Arch, Hyde Park Corner, London (English Heritage Photo Library).

Contents

Acknowledgements vi

Introduction
John Fidler vii

Part I: Research 1

Stone consolidants: Brethane™
Report of an 18-year review of Brethane™-treated sites
Bill Martin, David Mason, Jeanne Marie Teutonico, Sasha Chapman, Roy Butlin and Tim Yates 3

Lime treatments:
An overview of lime watering and shelter coating of friable historic limestone masonry
John Fidler 19

Lime method evaluation:
A survey of sites where lime-based conservation techniques have been employed to treat decaying limestone
Catherine Woolfitt 29

An investigation of sacrificial graffiti barriers for historic masonry
Nicola Ashurst, Sasha Chapman, Susan Macdonald, Roy Butlin and Matthew Murray 45

Soft wall capping experiments
Heather Viles, Chris Groves and Chris Wood 59

Part II: Development and Case Studies 75

Conserving fractured and detaching stone tracery:
Developing a technique for the stabilization and consolidation of fire-damaged tracery at the Church of Holy Cross Temple, Bristol
Chris Wood and Colin Burns 77

A solution for the stone repair of a cracked primary column at the Wellington Arch, Hyde Park Corner, London
Les Ayling 97

Acknowledgements

Current and past staff of English Heritage, former contracting scientists and current technical collaborators have all contributed to this long-awaited volume concerned with stone building materials, masonry construction and associated component systems and studies of their decay and treatment.

Two former staff of the Building Conservation and Research Team, Jeanne Marie Teutonico and David Mason, have both managed the Commission's building materials research programme and have been series editors of English Heritage's Research Transaction volumes. They both participated in Project AC2 Stone Consultants and are joint authors with fellow English Heritage conservators, and scientists from BRE, of a final report on the work presented here. English Heritage acknowledges that without their sterling efforts over several years, the current crop of technical and scientific reports would not have been so successfully released into the public domain.

Special mention should be made of Les Ayling, one of English Heritage's current conservation (structural) engineers who confessed that his case study of an ingenious masonry repair technique used on the recent conservation of the Wellington Arch in London was his first ever publication. Chris Wood continues to be the Building Conservation and Research Team's most prolific writer, with contributions to two papers in this volume. Thanks are also due to Professor John Ashurst, Nicola Ashurst, Susan Macdonald, Sasha Chapman and Colin Burns, all former colleagues, for their efforts on English Heritage's behalf in delivering past research and writing current reports.

Collaborating research contractors deserving credit at the Building Research Establishment, now formally known as BRE, include Dr Tim Yates and Matthew Murray. I am also grateful to Dr Roy Butlin (now retired) for his former management of English Heritage's research programme there. Grateful thanks are also due to Heather Viles and Chris Groves from Oxford University for their report of promising results from experiments on studies of soft wall capping.

The consultant editor for Volume 2 was Dr Kate Macdonald to whom English Heritage, as ever, is indebted for her assistance and patience in bringing together and structuring diverse papers from several authors over such a long gestation period.

Introduction

JOHN FIDLER
English Heritage, 23 Savile Row, London W1S 2ET, UK; Tel: +44 207 973 3025;
Fax: +44 207 973 3130; email: John.Fidler@english-heritage.org.uk

---

This long-awaited volume of English Heritage's *Research Transactions* is devoted to research and case studies on the repair and conservation of historic stone masonry. Some of the projects have been a long time in gestation (Martin 1996, 43–45) for important scientific reasons and are reported here at their conclusion. Other projects are at an interim stage and are described for the first time. Still others, principally case studies, are completely new, topical and previously unreported.

As one might expect from an organisation that promulgates a highly conservative position on stone conservation, in terms of ethics and techniques, English Heritage's view is that too much building work takes place that is either unnecessary or inappropriate. By thoughtless or ill-informed actions on the part of specifiers, conservators and contractors perfectly viable historic fabric is being lost before time and ineffectual treatments deployed that waste scarce resources. Collective assumptions about the permanence of certain remedial processes are unfounded. Too little time has been spent studying the real impacts of treatment cycles.

The papers that follow divide themselves into three components (research, development and case studies) and into two distinct sets of work, both concerned with the retention and extension of historic masonry's life.

- The first set focuses on systems of protection, through the use of organic and inorganic masonry consolidants; through wax defences in sacrificial graffiti barriers; and through the employment of soft wall cappings.
- The second set deals with keyhole or microsurgery techniques in situations where the large-scale and costly replacement of historic fabric would otherwise be necessary.

RESEARCH

Four of the research projects published here are concerned with technical studies of treatment impacts on historic masonry. In the case of the Brethane™ review, long-term field exposure trials which lasted for almost two decades were monitored, recorded and analysed. As for the lime treatment method used for the protection of limestone and plaster, one paper recaps on the technique and reviews previous inconclusive laboratory tests, while another describes a snapshot field survey of treated sites and the system's performance. In the case of the sacrificial graffiti barriers, the paper addresses accelerated multiple retreatment and removal trials, a recurring theme running through most of the work in this *Transactions* volume.

DEVELOPMENTAL WORK

The last of the research papers addresses the biological, physical and chemical functions of soft (vegetation) cappings for stone walls. Like stone consolidation with the lime water (calcium hydroxide) treatment, referred to above, soft wall cappings have taken on a mythological status amongst their proponents and are seen as a ubiquitous panacea for all manner of ills confronting ruined masonry protection. Like the lime water technique, nobody has successfully understood what, if anything, is being achieved through soft wall capping, though plenty of hypotheses remain untested.

The preliminary testing at Oxford University of the function of soft wall cappings (Ashurst & Dimes 1990, 6–10) and the quite separate masonry consolidation case study by English Heritage staff on Bristol Temple Church, show the usefulness of technical developmental work. At Oxford, the short-life pathfinder programme was designed from field observations to consider the various actions attributed to soft wall cappings by their advocates and to test whether scientific assessments could be achieved in facsimile laboratory conditions. Through such simple, cheap and effective work it has been possible to fine-tune the design, costing and programming of a more extensive testing regime that should get to the bottom of the technical problem and provide practical answers. In Bristol, the site trials and errors led to a workable technical solution that was tested and then promulgated to fellow specifiers, conservators and craftsmen for wider field application.

Brethane™ review: project AC2

In one sense, studies of Brethane™ are somewhat outmoded. Market forces have played their part in making this material relatively unpopular among professional conservators in the field in recent years, mostly because of its complicated installation, health risks during application and relative costs and availability. Yet its essential characteristics as a consolidant, once deployed, are directly comparable to the more popular materials in use today and so make reports of its performance both topical and relevant.

The impact studies of the alkoxysilane consolidant, Brethane™ (Price 1981) have been ongoing for many years and it has been difficult to decide when to call a halt and report. The longer the field trials, it is argued, the more accurate an assessment can be of material weathering characteristics and performance over time. However, as time progresses, so too do technical and scientific thought and practice, to the extent when the originating research protocol looks rudimentary by current standards. With the luxurious benefit of hindsight, we can now see that the research started in the late 1970s and early 1980s was much more subjective than we might have hoped.

In comparable though more recent work soon to be reported, we have learnt several lessons from our Brethane™ experience and have deployed a wide range of newly developed techniques to make such studies more objective for the benefit of posterity. For example, English Heritage's Building Conservation and Research team now regularly uses decay mapping techniques, combined with close range photogrammetry and rectified stereo photography to characterize deteriorating surfaces before intervening, so that historic and current rates of decay and loss can be assessed throughout the duration of the project. We also carry out detailed material analysis to evaluate salt and moisture conditions; we deploy environmental monitoring devices to ascertain ambient microclimates within which decay processes are active and our treatment trial has to perform.

Nonetheless, it was a farsighted and bold decision by English Heritage and its predecessors to develop a long-range field trial at all and the data being generated from it is bound to inform the complementary, though less realistic, laboratory-based accelerated weathering tests on treated cubes that manufacturers and others have employed for much shorter periods. If English Heritage cannot undertake long-term field trials, who can?

We must also remember the political climate within which studies of stone consolidants used to take place. At the start of the Brethane™ work, the Society for the Protection of Ancient Buildings called for a national moratorium on the use of such materials (SPAB 1980) fearing that the appearance of historic masonry would be permanently affected or that actual damage might ensue. Happily, English Heritage's relations with the Society have grown stronger over the years and differences over the use of consolidants are much less polarized.

The long-term field trials that the SPAB demanded in its leaflet continue to be monitored and now bring into focus some very important philosophical and technical questions. As predicted by English Heritage and Brethane's creator, the effect of the consolidating material has caused no long-term damage to the subject stonework and has slowed down decay and extended its life. Algaes, lichens and mosses, that provide so much of the colour and texture of historic masonry by contributing to its patination, were able to grow back over treated areas without problems. Thus the character, appearance and welfare of the decaying masonry have been enhanced.

Interestingly, the limestone sites appear to have benefited to a greater extent than the more chemically compatible sandstone sites. This leads one to reflect on the hydrophobic (water repellent) characteristics of this particular consolidant when used on soluble stones. How much of the benefit of consolidation can be attributed to water repellency, and how much to the consolidating or strengthening characteristics? Could an equally beneficial effect have been achieved by the use of a less long-lasting water repellent applied for preventative maintenance reasons? Did the consolidant's actions benefit from application to limestone's greater pore structure, where depth of penetration might be more easily achieved and the interlocking pores give better physical anchorage sites? Food for thought.

From this unique field assessment we now understand that the consolidant itself should not be called a 'preservative', which infers permanent preservation. It is found to decay and weathers back from the face of the masonry and so only slows the decay of the stone and does not permanently prevent it. Many of those involved in conservation have previously believed that treatment with consolidants was a once-only, perfect, final solution to stone decay. Such attitudes will now have to change.

If the consolidant decays, can its effective weathering surface be repaired, strengthened or buffered against deterioration? How could the application of additional new consolidant realistically be achieved? Surely its solvent delivery system would affect the existing material? Can one deploy an ethyl silicate stone consolidant over a completely different and now aged system? With what results for the consolidating process and performance? And more especially, with what consequences for the underlying stonework?

In addition, if stone consolidants in the longer term now have to be maintained or reapplied, what are the impacts of multiple retreatment in cycles repeated, every twenty years say, for hundreds of years? Fundamental debates now need to take place which go to the very heart of conservation ethics (SPAB's reticence is still relevant). Of course, English Heritage's general position towards stone consolidants has not fundamentally changed: they should remain treatments of last resort (Ashurst & Ashurst 1988, 89; Martin 1996, 48 & 49). But now the conservation community needs to realistically address the practicalities or otherwise of attaining the aims of the Venice Charter (ICOMOS 1966) with respect to these treatments. Can they ever be reversible and non-prejudicial to future interventions? The authors here postulate on possible ways ahead.

English Heritage is not resting either. Plans were recently formulated with international scientific collaborators, including the laboratories of the Soprintendenza per i Beni Artistici e Storici del Veneto, those of the Bavarian State Heritage Service and, in France, the Laboratoire du Centre Interregional de Conservation et Restauration du Patrimoine (Laboratoire du CICRP) and the Laboratoire de Recherche des Monuments Historiques (LRMH), and an application submitted to the European Commission Research Directorate for a grant under the Fifth Framework round, third call, to study the impacts of masonry consolidant retreatment on historic masonry. This application was unfortunately rejected, and the chances of reapplying under the forthcoming Sixth Framework, due to a change of criteria, look bleak. However, English Heritage is in discussions with the Getty Conservation Institute on how this important international work may be carried forward.

Lime treatments: project AC9

English Heritage has argued for some time (Fidler 1995, 55; Martin 1996, 39) that more research is necessary to determine if and how the combination of lime treatments actually works. Perhaps the laboratory tests deployed by the Building Research Establishment (now BRE) thirteen and more years ago were too aggressive or inappropriate to measure any tangible effects of inorganic consolidation? Certainly, international knowledge on crystal growth, testing regimes and other salient factors has been rapidly developing in the last few years. With this in mind, English Heritage postponed the start of its laboratory work and instead, under the auspices of its project AC9, invested in a review of past scientific work besides commissioning a field assessment of previous treatment applications and their durability, both reported here.

In December 2000, the draft papers informed a joint scientific seminar which English Heritage hosted with the Getty Conservation Institute at the Society of Antiquaries in London. The output from this intensive two-day workshop on progress on inorganic masonry consolidants is being prepared for submission for peer review and publication in the International Institute for Conservation's new 'Reviews in Conservation' series. English Heritage and the Getty Conservation Institute are interested in forming research consortia now to take forward a plan of work based on this review.

Sacrificial graffiti barriers: project AC7

This research paper reports on extensive trials with waxed-based sacrificial graffiti barriers on brick and limestone masonry which took place at BRE in the early and mid 1990s. The design of the tests predates subsequent work by BRE for Historic Scotland, mostly on the barriers' effects on sandstones (Historic Scotland 1999), and is post-dated by the publication of English Heritage's own free technical advisory note on the subject (English Heritage, 1999). Delays in publication were caused by numerous personnel changes within English Heritage's Building Conservation and Research team and at BRE, and also because the final output needed to take account of any relevant changes emanating from the revised code of practice on cleaning buildings (BSI 2000).

Permanent, varnish-type, surface coatings to limit graffiti damage have long been dismissed by conservators for use on historic masonry because of the dramatic alterations to the stone's surface appearance and the lack of realistic reversibility, but also due to risks from the likely entrapment of subsurface salts. Wax-based systems offered the possibility of being able to remove the treatment after every paint attack. Sadly for building conservation, the research findings reported here show that there appears still to be a very wide gap between technical theory, marketing claims and actual practice.

Soft wall cappings: project AC26

Advocates of soft wall cappings cite their use on English ruins from Hadrian's Wall [1] to Chysauster [2] as reasons why the 'technique' should be adopted more widely to preserve exposed decaying stone wall tops. Of course, it can be debated whether the remnant vegetation at these sites was purposefully planted with conservation intent, or reflects previous neglect and current respect for the natural heritage. After all, several English Heritage sites in guardianship which come down to us as masonry ruins are now not only Scheduled Ancient Monuments but also Sites of Special Scientific Interest (SSSI), preserved as much for the flora and fauna inhabiting the old stonework, as for the structure itself. [3]

The United Kingdom is renowned the world over for the 'style' of its care of ancient monuments (Thompson 1981). With plan forms delineated by gravel floored rooms and grassy exteriors, and walls 'frozen in time' with their leaning, bulging deformities, the Ministry of Works' techniques, developed from Ruskinian principles at the turn of the twentieth century, made sure that four hundred years of neglect and decay were curtailed, for the most part by eradicating the highly damaging vegetation, notably ivy.

Now there is a softening of this centralized approach to site presentation with a return to a more natural, picturesque appearance in one or two cases, made affordable these days in conservation and economic terms through improved maintenance management. Sites such as Wigmore Castle [4] exemplify this approach but are only following tenets laid down at the end of the Victorian era in the repair of urban monuments such as Birkenhead Priory (Chitty 1987, 57 & 58).

The real question is, should such treatments be encouraged? How can national policy be set without truly understanding whether and how soft cappings affect the monuments in our care? This is where technical research can play its part in making such decisions.

The report in this volume of the Research Transactions is of work in progress on this subject. Tests were devised by the Built Environment Research Group in the Department of Geography and the Environment at Oxford University to investigate the thermal blanketing effect of soil and vegetation on soft stone wall heads of ruined Scheduled Ancient Monuments. The hydrological impacts of such coverings were also studied and the results designed to refine the experiments for longer-term trials.

Clearly there is more work to do to create realistic simulations of field conditions and studies of a variety of different soils and vegetation would account for the range of covering techniques found around the country. But the research shows how the cappings might, in some cases, offer a degree of physical, chemical, temperature and humidity buffering to a fragile substrate.

CASEWORK REPORTS

Bristol Temple Church stone consolidation

Because of the unique forms of masonry deterioration caused by a Second World War fire bomb, the resultant conflagration and the rapid cooling and shattering by fire-fighting water, Bristol Temple Church has been a test bed for many unusual, pioneering attempts at conservation, all with their successes and failures. Until the present time few have been sufficiently well recorded to be published but also to be monitored as to their future performance. The paper presented here reviews past work on the site and describes new, sensitive ways to pin, grout and consolidate broken masonry.

After war damage, the shattered stone tracery of the church's aisle fenestration was barely kept in place by semi-permanent and highly disfiguring timber corsets bolted around the window geometry. In the 1980s (Ashurst & Dimes 1990, 47), a more enlightened substitute, lightweight flexible glass fibre rods, was inserted into pre-drilled holes and locked in place by epoxy resins to provide concealed structural support. Phosphor bronze wire was then threaded around the rods to create an armature and the whole made up to profile in lime-based mortar.

But time has shown this type of repair to have been too rigid for the long-term welfare of the ruin. So well designed was the treatment however that it failed first, as intended, so as not to jeopardize the monument itself. Now an ingenious new technique has been deployed which is based on even lighter lightweight micro-stitching with copper cables, held in more flexible grout. The concept is designed to permit the large number of shattered stone elements to react to thermal and other movement forces without losing overall structural integrity.

Wellington Arch stone repairs

Finally, the short paper on recent masonry repairs to a cracked column at the Wellington Arch, Hyde Park Corner, London, highlights the ingenious adoption of carpentry repair or metal-stitching techniques to stonework. This has only been made possible by new developments in stone-cutting technology, which, when combined with good craftsmanship, are more than a match for many of the more unusual structural masonry problems we face.

ENDNOTES

1 Hadrian's Wall is a linear Roman fortification in the north of England, now a Scheduled Ancient Monument and World Heritage site, stretching 117 km (73 miles) from the Solway Firth in the west to the Tyne in the east and is easily reached by train, bus and car from either Carlisle or Newcastle, and places in between. Its western third is made up in part of stone and earthen banks with grassed tops.

2 Chysauster ancient village (OS map 203; SW473350), 5.6 km (3.5 miles) from Penzance, is a deserted Romano-Cornish village with a 'street' of eight well-preserved houses, each comprising a number of rooms around an open courtyard. Standing, coursed rubble, dry stone walls are between 1.5 and 2 m high and topped with thin soil, luscious grass and mosses.

3 Rochester Castle (OS map 178; TQ742686) lies adjacent to the town bridge across the River Medway in Kent. Parts of its thirteenth-century curtain walls are designated as an SSSI due to the pockets of rare flowers, French Pinks, whose seeds came across with the original masonry from Normandy.

4 Wigmore Castle (OS map 137; S040906918), lies eight miles west of Ludlow in Shropshire on the A4110 about 1.2 km (0.75 miles) from the village centre by footpath. It is a thirteenth- and fourteenth-century castle which was dismantled in the Civil War and remained abandoned and neglected until 1998–9 when partial clearance of vegetation and consolidation of walls took place under English Heritage's direction. During this process, many of the ferns and other 'soft' plants adorning the ruined masonry were temporarily removed, stored and reused after the stonework had received major structural repairs and consolidation.

BIBLIOGRAPHY

Ashurst J and Ashurst N, 1988 *Stone Masonry*, Practical Building Conservation, English Heritage Technical Handbook series. Vol **1**, Aldershot, Gower Technical Press.

Ashurst J and Dimes F G, 1990 *The Conservation of Building and Decorative Stone*. 2 vols, Oxford, Butterworth-Heinemann.

British Standards Institution, 2000 *The Cleaning and Surface Repair of Buildings BS Code of Practice 8221/2* [formerly BS 6270: 1982] London, BSI.

Chitty G, 1987 A prospect of ruins, in *ASCHB Transactions*, **12**, 43–60.

English Heritage, 1999 *Graffiti on Historic Buildings and Monuments: Methods of Removal and Prevention*, Technical Advisory Note, London, English Heritage.

Fidler J, 1995 Lime treatments: Lime watering and shelter coating of friable historic masonry, in *APT Bulletin*, **XXVI**: 4, 50–56.

Historic Scotland, 1999 *The Treatment of Graffiti on Historic Surfaces: Advice on Graffiti Removal Procedures, Anti-Graffiti Coatings and Alternative Strategies*, Technical Advisory Note 18, Edinburgh, Historic Scotland.

International Council on Monuments and Sites (ICOMOS), 1966 *The Venice Charter 1994*, Paris, ICOMOS.

Martin W, 1996 Stone consolidants – a review, in *A Future for the Past: Strategic Technical Research in the Cathedrals Grants Programme, Proceedings of a Joint Conference of English Heritage and the Cathedral Architects Association held in London 25th & 26th March 1994*, London, James & James (Science) Publishers Ltd, 30–50.

Price C, 1981 *Brethane Stone Preservative. Current Paper 1/81 January*, Garston, Building Research Establishment & DoE.

Society for the Protection of Ancient Buildings, 1980 *The Development of Silane-Based Preservatives: Their Use and Abuses*, London, SPAB.

Thompson M W, 1981 *Ruins: Their Preservation and Display*, London, British Museum Publications.

ADDRESSES

BRE (formerly known as the Building Research Establishment), Garston, Watford WD2 7JR, UK

Getty Conservation Institute, The J P Getty Trust, Suite 700, 1200 Getty Centre Drive, Los Angeles, CA 90049-1684, USA

International Institute for Conservation of Historic and Artistic Works (IIC), 6 Buckingham Street, London WC2N 6BA, UK; www.iiconservation.org

AUTHOR BIOGRAPHY

John Fidler RIBA is a chartered architect and Head of Building Conservation and Research at English Heritage, where he is responsible for developing technical policy and providing technical advice for research and development work on building materials decay and their treatment, for technical training and for the organisation's outreach campaigns and technical publications. He was the first Historic Buildings Architect for the City of London Corporation, the first national Conservation Officer for Buildings at Risk, and was the youngest and last Superintending Architect to maintain the country's historic estate. Fidler is the author of numerous technical publications on aspects of building conservation and was for ten years architectural editor of *Traditional Homes* magazine.

Part I
Research

Stone consolidants: Brethane™
Report of an 18-year review of Brethane™-treated sites

BILL MARTIN [*], DAVID MASON, JEANNE MARIE TEUTONICO AND SASHA CHAPMAN
English Heritage, 23 Savile Row, London W1S 2ET, UK; Tel: +44 207 973 3073;
Fax: +44 207 973 3130; email: william.martin@english-heritage.org.uk
ROY BUTLIN AND TIM YATES
Building Research Establishment, Garston, Watford WD2 7JR, UK

Abstract

This paper summarizes the results of an 18-year project to monitor the results of consolidation trials using the alkoxysilane consolidant Brethane™ at a number of English Heritage sites. This long-term study has provided an opportunity to evaluate the performance of this consolidant on different stone types in a range of conditions. The results provide evidence that this consolidant may help to slow down the rate of decay of many stone types, although effectiveness of the system diminishes over time as the consolidant itself degrades. Unexpectedly, observations indicate that Brethane™ appears to work better in limestones than in sandstones, which are composed of material more chemically similar to it.

Key words

Stone consolidants, Brethane™, water repellency, survey methods

INTRODUCTION

Brethane™ is a three-component consolidant system for stone based on methyltrimethoxysilane (MTMOS). Developed at the Building Research Establishment (BRE) during the 1970s, it was designed as an alternative to uncatalyzed silane systems. Uncatalyzed silanes, which cure slowly by a combination of hydrolysis and condensation, have been found to deposit eventually less than 40% of the total volume of resin applied, due to loss of impregnation material through evaporation of the solvent during the polymerization process.

Brethane™, which has been marketed since 1983 under a British patent, is supplied as a three-pack system comprising a trimethoxymethylsilane, an organo-metallic catalyst (lead carboxylate), and solvent (ethanol). The components are mixed immediately before application, which is usually by hand spray. Upon mixing, a thin colourless liquid is produced, a silane monomer quickly absorbed by the porous stone to which it is applied. The resin generally penetrates to a depth of 25–50 mm (1–2 ins), over a period of two to four hours, before setting rapidly to a gelled state. A number of factors control the actual depth and pattern of penetration, including the porosity of the stone, its pore-size distribution, the angle of inclination of the surface and the structure and alignment of bedding planes. The rate of set may to some extent be controlled by varying the strength of the catalyst. This is supplied in two different grades, and an appropriate strength is chosen depending on the ambient temperature at the time of application. The rate can be monitored during application with reference to small control bottles, filled with a quantity of resin at the time of mixing. Upon initial polymerization, the gelled Brethane™™ shrinks back, over a period of days, to form a network of silica on the exposed surface and on the internal pore walls, binding fragile decayed stone to underlying sound stone, and at the same time imparting a degree of water repellency.

DEVELOPMENT AND USE OF BRETHANE™

In the process of developing Brethane™, extensive trials were carried out to examine the consolidating potential of suitable chemicals, and the effect of such chemicals on stonework. A number of reports were produced at the time (BRE 1976a, BRE 1976b, BRE 1978, BRE 1981a, BRE 1981b, BRE 1983)

Subsequently two further papers were produced. The first reported the results of a study of the depth of penetration of Brethane™ using infrared spectrometry (Ridal 1988), and the other, the results from the long-term exposure trial of stone blocks treated with Brethane™ and other consolidants. (Butlin et al 1991)

During its early development Brethane™ was not commercially available. However, the system was tested in the field during the late 1970s and early 1980s. These trials were conducted by the Research and Technical Advisory Service of the Directorate of Ancient Monuments and Historic Buildings of the Department of the Environment (subsequently called the Historic Buildings and Monuments Commission for England, colloquially known as English Heritage from 1984). Detailed records were made of all treatments applied by the organisation's direct labour force at some 54 sites. These early experiences were intended to support laboratory tests, and to substantiate and inform the development of the system through an extensive programme of on-site monitoring, and early results were promising. The product began to be manufactured and marketed commercially, under licence (by Colebrand Ltd[5]), in 1983, together with a users' handbook including pro forma report sheets for

[*] Author for correspondence

recording the details of any treatment. Until 1990 all operatives intending to apply Brethane™ were required to attend a two-day training course. Since 1983, treatment records have continued to be maintained, in the majority of cases, by the supplier, providing a vital body of documentary material for inclusion in a comprehensive register of treated sites now established by English Heritage.

The maintenance of detailed records was regarded as an essential aspect of the development programme. Even more importantly, however, the need was identified at an early stage for a systematic programme of regular inspection and monitoring so that the long-term effectiveness of Brethane™ could be evaluated. Consequently, Brethane™ has been subjected to a unique field monitoring exercise over an 18-year period, beginning in 1980, carried out by the Architectural Conservation Team (now Building Conservation and Research Team) at English Heritage, focused on a selection of the 54 sites. The results are reported below.

BACKGROUND TO THE ENGLISH HERITAGE TRIALS

At each of the 54 sites treated between 1976 and 1983, a decision had been made to use Brethane™ because a perceived problem of stone deterioration had come to light, and it was felt that Brethane™ might offer a potential solution, and have advantages over other systems available at the time. In some instances the treatment formed part of a more comprehensive package of conservation work, whereas in others it took the form of a test on a limited area of stonework. The 54 sites were not pre-selected as representative of building typology, geographic location, petrology or because they exhibited a characteristic range of decay phenomena. The decision to use Brethane™ was driven by practical necessity.

In 1984, 10 sites were selected from the total of 54 in the UK for inclusion in a projected long-term field evaluation programme. This was the first time that such an extensive, real-time study of consolidant performance had been undertaken. The ten sites identified were:

- George III Temple, Audley End House, Essex
- Berry Pomeroy Castle, Devon
- Bolsover Castle, Derbyshire
- Chichester Cathedral, West Sussex
- Goodrich Castle, Hereford and Worcestershire
- Howden Minster, North Yorkshire
- Kenilworth Castle, Warwickshire
- Rievaulx Abbey, North Yorkshire
- Sandbach Crosses, Cheshire
- Tintern Abbey, Gloucestershire

These sites were chosen as a cross-section partly because they exhibited the most typical and representative range of stone deterioration problems in the UK. In addition, however, it was considered that reliability of records, accompanied if possible by post-treatment documentation, should influence the selection of sites for the programme. Records of the initial Brethane™ treatment existed for all of the 54 cases (see Annex A). Only seven sites, however, had been revisited on a regular basis, generally some years after the initial treatment had taken place, as part of an *ad hoc* evaluation exercise to assess the performance of the system. Records for these sites therefore existed, detailing the observations made during (in some cases) more than one post-treatment survey. To these seven sites, three new sites were added, all of which had been treated with Brethane™ no more than 12 months previously.

While petrological analysis had been carried out on stone samples taken from the treated area at the majority of sites, it was not possible to do this at all ten sites. Therefore geological descriptions and porosity readings for Audley End, Chichester Cathedral, Bolsover Castle SW Entrance and Sandbach Crosses are based on data available, from a number of sources, from stones of the same geological age and a similar petrological character.

SURVEY METHODOLOGY

It was apparent that a practical and readily understandable system of recording certain agreed surface phenomena would need to be developed for the survey programme. This system would need to take into account the nature of available recording technologies, as well as the ability of future surveyors to repeat the survey programme without the need for elaborate training, equipment or resources. Given the nature of the sites, the range of materials involved, the proposed time-scale, the budgetary restrictions imposed and the experience of the personnel involved, it was decided to use a simple system, relying largely on sight and touch. In each case, an area of treated stonework would be compared with an adjacent untreated or control area, and assessed visually and by touch, against a set of pre-established criteria, during each visit to the site. This approach was modelled on that used by the Ancient Monuments Branch of the Ministry of Works, and devised jointly by the Branch and the Building Research Establishment, for recording the effectiveness of stone preservative treatments in the 1960s (Clarke & Ashurst 1972)

Seven assessment criteria were defined, each one with a letter code. Each criterion was to have its own numerical value scale, so that a performance 'marking' for each criterion could be awarded each time the treated area was inspected (Table 1). The seven criteria were grouped into three categories:

- appearance of treated stonework (colour, degree of soiling, biological decay, decay relative to untreated stone)
- condition of stonework (degree of powdering and scaling)
- water repellency.

All values were based on a visual, and, to some extent, tactile assessment of surface appearance and condition, with the exception of the water repellency test, which

Table 1. *Key to the scoring and codes in Tables 2–12.*

Appearance of treated stonework
AA 1 Lighter than untreated stonework
    2 Much the same as untreated stonework
    3 Darker than untreated stonework
AB 1 Cleaner than untreated stonework; streaked
    2 Uniformly cleaner than untreated stonework
    3 Much the same as untreated stonework
    4 Uniformly dirtier than untreated stonework
    5 Dirtier than untreated stonework; streaked
AC 0 No biological growth
    1 Less biological growth than untreated stonework
    2 Same biological growth as untreated stonework
    3 More biological growth than untreated stonework
AD 1 Generally less decayed than untreated stonework
    2 Much the same as untreated stonework
    3 Generally more decayed than untreated stonework
Condition of stonework
BA 0 No powder transferred to finger drawn lightly across surface
    1 Slight amount of powder transferred to finger drawn lightly across surface
    2 Substantial amount of powder transferred to finger drawn lightly across surface
    3 Pieces of stone fall away when finger drawn lightly across surface
BB 0 No scales present
    1 Surface scales present, less than 10mm across
    2 Surface scales present, greater than 10mm across
    3 Surface cracking and crazing
Water repellency of treated stonework
CA 0 Does not absorb water; water sits on surface in globules or runs straight off
    1 Does not absorb water; film of water remains on surface
    2 Absorbs water less readily than untreated stonework
    3 Absorbs water as readily as untreated stonework

Limestones of this type may have a porosity measurement of up to 25%. In the nineteenth century the entablature was replaced in Coade stone (a form of artificial stone).

Two of the columns have been used for a series of consolidant trials. In 1973, during the early development work for Brethane™, different areas of stonework on the was carried out using a dropper to apply water to treated and untreated surfaces, the effect being noted by visual comparison.

A set of pre-determined 'acceptable values' was defined for each of the seven criteria. At the end of the survey period, the performance of the Brethane™ treatment in relation to the 'acceptable value' for each assessment criterion could be evaluated. Where the performance marking for any of the measurement criteria in each of the surveys undertaken was seen to be equal to or less than the acceptable value, then the treatment could be regarded, in relation to that particular criterion, as successful. Where the overall performance marking exceeded the acceptable value, then, in relation to that criterion, the treatment could not be regarded as successful.

The results for each site are given in Tables 2–12. Comments on individual sites and the interpretation of the observations are given below.

The surveys, recorded on pro forma record sheets (see Annex B), were carried out by personnel from English Heritage or from the BRE. Given the timescale of the monitoring, it was perhaps inevitable that personnel changes in both organisations would occur. It was recognised that this, and the reliance in the survey system on visual observation, might have an effect on the objectivity of the survey.

THE SITES

Audley End House (Figures 1–2, Table 2)

Audley End House, Essex, is a Jacobean mansion standing in extensive parkland, in which are situated a number of monuments and garden buildings. The George III Temple, built in the eighteenth century, comprises an entablature with inscription and Corinthian columns in clunch, a greyish-white limestone from the Cretaceous Lower Chalk.

Figure 1. *General view of the George III Temple at Audley End House (English Heritage Photo Library).*

Figure 2. Detail of Column 5 at the George III Temple showing the area treated with Brethane (1997) (English Heritage Photo Library).

Table 2. Scoring table for Audley End, Panel E (treated with Brethane in 1976) limestone (clunch).

	01/80	08/83	12/84	02/86	05/88	12/91	04/96	acceptable values
AA	1	1	1	1	2	1	2	2
AB	2	2	2	2	3	3	3	<4
AC	0	0	0	0	0	0	0	<2
AD	1	1	1	1	1	1	1	<2
BA	1	1	1	1	2	1	1	<2
BB	1	1	1	1	1	1	1	0
CA	1	1	1	1	1–2	2	2	<3

columns were treated with silane (Area B), and an acidified silane (Area A).[1] In 1976 three areas were treated with Brethane™ (Areas D, E and F). One of these (Area D) was then immediately retreated, and Area B, which had been treated with silane in 1973, was also treated with Brethane™. Area C was treated with the stone preservative Consolith.[2]

Overall, results of the treatment at Audley End over a 20 year period are good, and the system's performance is exceptionally stable. There is little difference between the condition of treated stone in 1980, and its condition in 1996. Although water repellency appeared to have decreased after about 12 years, when last examined treated areas still exhibited a significantly greater degree of water repellency over untreated areas. The area that had been treated twice with Brethane™ showed no evidence of any problems arising because of this. This area has remained much less decayed than the adjacent untreated areas. The most recent visit took place in 1996 and there appeared to be little change in the condition of the stone since the previous visit.

Berry Pomeroy Castle (Figures 3–4, Table 3)

Berry Pomeroy Castle, Devon, the remains of a late medieval fortified structure, is situated about nine kilometres (5.5 miles) west of Torquay. The treated stone is an off-white, chalky limestone, shown by petrographic and thin section analysis to be a poorly washed biosparite. The porosity of the specimen analysed was patchy, but moderately high. It is typical of the building stones quarried from the Middle Chalk and is believed to be either Beer stone, the main source of which is Beer, near Axminster, or one of the local lateral equivalents.

The base of one of the piers of the loggia was treated with Brethane™ in 1983. The pier is composed of carved facing stone and a rubble core. At the time of treatment, there were doubts about the chances of success of the consolidant because of the low uptake of Brethane™ (3.7 litres/m^2 compared to an expected 5.0 litres/m^2, or 0.72pts/ft^2 to 0.97pts/ft^2). However, when last surveyed in 1997 the treated stonework was found to be in reasonably sound condition. This is probably a result of the combined effect of the treatment, and the fact that the stonework at this site is protected by a wooden box during the winter months. However, surface cracking was evident in areas where physical damage and previous stone loss had occurred.

Overall, Brethane™ treatment effectively improved water repellency and had a good consolidative effect in the first few years. After five to ten years, treated stone was found to be in much the same condition as untreated stone, and water repellency was beginning to diminish, though still greater than in control areas.

Table 3. Scoring table for Berry Pomeroy Castle (treated with Brethane in 1983, limestone).

	08/85	06/86	02/88	02/92	08/97	acceptable values
AA	1	1	2	1	1–2	2
AB	2	2	1	2	3	<4
AC	0	0	1	0–1	1–2	<2
AD	1	1	2	1	2	<2
BA	0	0	0	0	0–1	<2
BB	0	0	1–3	1	3	0
CA	1	1	1	1–2	1–2	<3

Table 4. Scoring table for Bolsover Castle, SW Turret (treated with Brethane in 1982, magnesian limestone).

	08/83	03/84	04/86	03/88	02/92	06/97	acceptable values
AA	3	1	1	1	1	2	2
AB	2	2	2	3	3	3	<4
AC	0	0	0	3	1	2	<2
AD	1	1	1	1–2	1	1–2	<2
BA	0	0	0	0–1	1	1–2	<2
BB	0	0	0	2–3	3	1–2	0
CA	0	0	0	0	2	2	<3

Figure 3. Base of the loggia pier at Berry Pomeroy Castle, treated in 1983 (1997) (English Heritage Photo Library).

Figure 4. Detail of the base of the loggia pier at Berry Pomeroy Castle (1997) (English Heritage Photo Library). See Colour Plate 1.

Bolsover Castle, SW Turret (Table 4)

Bolsover Castle, Derbyshire, is an early seventeenth-century mansion on the site of a Norman castle, west of Bolsover, and approximately nine kilometres (5.5 miles) east of Chesterfield. A substantial area of the SW turret of the Little Keep was treated with Brethane™ in 1982. The turret is built of pale grey-orange dolomite (similar in appearance to the local Red Mansfield), with paler dolomitic sandstone used for the doorway and window dressings.

A stone sample from the site suggests that the treated stone is a siliceous dolomite from the Permian magnesian limestone, probably from a local quarry. Although consisting mainly of dolomite crystals (80%), the rock is relatively rich in quartz together with some fragments of siliceous rock (18%), and calcite (2%). The porosity of the weathered sample is high, being about 20%.

The Brethane™ treatment seems to have been generally successful, although powdering and localized scaling of surface layers were in evidence by 1997. After 15 years, however, general surface condition and water repellency were better than for untreated stone.

Bolsover Castle, SW Entrance (Figs 5–6, Table 5)

A die-stone in Hardwick sandstone, in the balustrade on the steps at the SW entrance to the Little Keep, was treated with Brethane™ in 1982. The stone is a very fine-grained lithic sandstone from the Middle Coal Measures, comprising mainly quartz and rock fragments, and is banded with reddish veins due to the variable distribution of iron minerals. No petrological analysis was undertaken, but sandstones from the Hardwick quarries are known to be of low porosity, typically 4.5% on average.

Examination of the monitoring records shows that the treated area was stabilized immediately after treatment, and remained fairly well consolidated thereafter, with slight powdering at the surface throughout the duration of the programme. Some cracking/crazing and a decrease

Figure 5. Treated block of Hardwick sandstone at Bolsover Castle, SW entrance, treated in 1982 (English Heritage Photo Library). See Colour Plate 2.

in water repellency have been noted in recent years, but overall condition is better than that of untreated stone, and there has been no permanent darkening.

Chichester Cathedral (Table 6)

The original construction dates from the eleventh century, with major thirteenth- and fifteenth-century additions, and some nineteenth-century rebuilding. Three blocks in the central mullion between the two-light openings in the Lantern Tower, (second, fourth and fifth courses from the sill on the north-facing internal reveal), were treated in 1978. The treated blocks are of Tisbury limestone, a glauconitic, siliceous limestone from the Vale of Wardour. No samples were analysed but stones

Figure 6. Untreated block of Hardwick sandstone at Bolsover Castle, SW entrance (English Heritage Photo Library). See Colour Plate 3.

Table 5. Scoring table for Bolsover Castle, SW entrance (treated with Brethane in 1982, sandstone).

	08/83	03/84	04/86	03/88	02/92	06/97	acceptable values
AA	3	3	3	2	3	3	2
AB	2	2	2	3	2	3	<4
AC	0	0	0	3	1	2	<2
AD	1	1	1	2	1	1	<2
BA	0	1	1	0–1	1	1	<2
BB	0	0	0	3	0	3	0
CA	0	0	0	1	2	2	<3

Table 6. Scoring table for Chichester Cathedral (treated with Brethane in 1978, limestone).

	08/80	07/83	06/84	06/86	04/88	05/92	03/96	08/97	acceptable values
AA	3	2	2	2	2	2	2	1	2
AB	3	3	3	3	3	3	3	2	<4
AC	0	0	0	0	0	0	0	0	<2
AD	1	2	1	1	2	1	1	1	<2
BA	1	1	1	1	1	1	1	1	<2
BB	0	0	0	0	0	0	0	0	0
CA	1	1	1	0	3	3	3	3	<3

Table 7. Scoring table for Goodrich Castle (treated with Brethane in 1978, sandstone).

	03/80	01/82	09/83	03/86	02/88	02/92	08/97	acceptable values
AA	2	2	2	2	2	2	2	2
AB	3	3	3	3	4	3–4	3	<4
AC	0	1	0–1	2	2–3	2–3	2–3	<2
AD	1	1	1	–	1–2	1–2	1–2	<2
BA	0	1	1	1	0–1	0	0–1	<2
BB	0	0	0	0	1–2	1–2	1–2	0
CA	1	1	1	1	2	3	2–0	<3

Figure 7. The 'Solar Column' at Goodrich Castle, partially treated with Brethane in 1978 (1997) (English Heritage Photo Library).

Figure 8. Detail of the 'Solar Column' (English Heritage Photo Library).

from similar beds quarried at Chilmark typically display a porosity reading of 3–5%.

Though initial darkening occurred, this faded and the treated stone has generally remained less deteriorated than untreated stone throughout the survey, with little or no sign of powdering or scaling. Water repellency was much improved at first, but declined after six to eight years, and is now no better than for untreated stone.

Goodrich Castle (Figures 7–8, Table 7)

Goodrich Castle, Herefordshire, is a twelfth- to fourteenth-century castle about four kilometres (2.5 miles) south of Ross-on-Wye. Hand samples indicated that the stones used for building are local calcareous sandstones, probably from the Devonian Old Red Sandstone, varying in colour from pale red-brown to green-yellow. They consist typically of mainly quartz grains in a calcite matrix, or in a mixed matrix of clay minerals and calcite. Porosity of two hand samples was shown to be low, probably less than 5% (macro-porosity).

Part of the central column supporting a cross wall in the 'solar' at Goodrich Castle was treated in 1978. The treated column appears to have been buried to about half its height when the castle fell into disuse.

The treatment appeared to have been successful when first applied, but by 1997 the condition of the treated area was found to be only slightly better than that of stonework in adjacent control areas, with extensive, local scaling having developed after ten years. The low porosity and low saturation coefficient (about 0.65) of the stone may explain why the treatment seems to have been less successful than in other cases (see Conclusions). Water repellency, though initially much improved, has begun to diminish in some areas, but has remained very high in others.

Howden Minster (Figures 9–10, Table 8)

Howden Minster, in the East Riding of Yorkshire, is a partly-ruined fourteenth-century minster some 36 kilometres (22 $\frac{1}{3}$ miles) west of Kingston-upon-Hull. Analysis of a hand sample has shown the stone at Howden to be a uniform, fine-grained dolomite, identified as silt-grade magnesian limestone, probably of Permian age, and likely to have come from quarries located about 30 kilometres (18 $\frac{1}{2}$ miles) to the west, or from the Tadcaster area. The porosity is 16%.

Two areas of the ruins at Howden Minster were treated with Brethane™. The first trial areas, covering parts of the internal wall arcades in the Chapter House, were treated in 1981. The second was a 1m² (9 ft²) area of walling to the exterior of the Choir, treated in 1984. The second of the two trial areas was treated as part of a

Figure 9. Howden Minster (English Heritage Photo Library).

Figure 10. South wall of the ruined choir at Howden Minster, showing the area treated with Brethane as part of a comparative study of three systems in 1984 (1997) (English Heritage Photo Library).

comparative study of Brethane™ and two other systems (the ethyl silicate Wacker OH,[3] and Belzona Clear Cladding, a silicone-based, colourless, surface water repellent in a white spirit carrier).[4] The condition of treated stone in this area was monitored at intervals between 1984 and 1997.

In 1997 the second of these Brethane™-treated areas was found to be very similar in condition to the control area, and showed similar levels of soiling to adjacent, untreated areas. Little evidence of significant or rapid recent decay, however, has been observed on any of the treated areas, or on the control areas.

Brethane™ appears to have improved water repellency of the treated area, while areas treated with alternative systems continue to absorb water as readily as untreated stone. As regards consolidation, there is now little difference between treated and untreated areas. There is no evidence that either of the other treatments in the trial have performed significantly better or worse than Brethane™. However, since decay generally in this part of the site is slow, it is difficult to gauge the effectiveness of the consolidants.

Kenilworth Castle (Figures 11–12, Table 9)

Kenilworth Castle, Warwickshire, is a twelfth-century castle just west of Kenilworth. The stone is a fine-grained sandstone, reddish-brown in colour, probably obtained from an adjacent quarry. Its composition is mainly quartz (around 64%), with rock fragments, and traces of biotite mica, feldspar and iron minerals. The primary cement appears to have been silica, though other materials are also present. A sample studied in thin section suggested a high porosity (20%).

Table 8. Scoring table for Howden Minster (treated with Brethane in 1984, magnesian limestone).

	02/85	11/85	04/86	11/86	12/87	05/92	06/97	acceptable values
AA	3	2	2	2	2	3	2	2
AB	3	2	2	2	3	4	3	<4
AC	0	0	0	0–1	2	2	2	<2
AD	1	1	1	1	2	2	2	<2
BA	0	1	0	0	0	1	1	<2
BB	0	0	0	0	0	0	3	0
CA	1	1	1	1	1–2	1–2	2	<3

Table 9. Scoring table for Kenilworth Castle (treated with Brethane in 1984, sandstone).

	09/84	02/85	02/86	10/86	02/88	05/92	06/97	acceptable values
AA	2	2	1	1	1	2	2	2
AB	3	3	3	3	3	3	3	<4
AC	0	0	0	0	4	0	1	<2
AD	2	2	1	1	1	2	2	<2
BA	2/3	2–3	2	2	2–3	0–2	1–2	<2
BB	0	0	0	0	1	1	0	0
CA	0	1	1	1	2	2	2	<3

Figure 11. Detail of the trial area of the south wall of the keep at Kenilworth Castle, showing the area treated with Brethane as part of a comparative study of three systems in 1984 (1997) (English Heritage Photo Library).

This site was also part of a comparative trial of Brethane™ and two other consolidants (Wacker OH and Belzona Clear Cladding). Three areas of stonework on the south wall of the Keep were treated in 1984, one with each of the three systems.

Following the Brethane™ treatment, the stone has remained, at best, only slightly better in condition than the control area, with some surface disaggregation, and very slight white bloom on lower areas. The ethyl silicate consolidant system (Wacker OH) used on the second panel seems to have had a comparable level of success in terms of consolidation, though with more extensive bloom. The treatment used on the third panel (Belzona Clear Cladding) has also been of limited effect. Here, too, powdering of the surface is in evidence.

Table 10. Scoring table for Rievaulx Abbey (treated with Brethane in 1979, sandstone).

	09/80	08/83	03/84	05/85	12/87	05/92	06/97	acceptable values
AA	2	1	1	1	1	1	2	2
AB	2	2	2	2	2	2	2	<4
AC	0	0	0	0	1	1	1	<2
AD	1	1	1	1	1	2	1	<2
BA	1	1	1	1	2	1	-	<2
BB	0	0	0	1	1–2	1–2	1–2	0
CA	1	1	2	1	2–3	3	-	<3

Figure 12. Efflorescence on the surface of treated areas at Kenilworth (1997) (English Heritage Photo Library). See Colour Plate 4.

Brethane™-treated areas do appear to have retained a higher level of water repellency than the other treated areas, but overall, at the end of 13 years, disaggregation has not been arrested, and Brethane™ cannot be said to have significantly outperformed the other systems as regards surface consolidation.

Rievaulx Abbey (Figures 13–14, Table 10)

Rievaulx Abbey, North Yorkshire, is a twelfth-century Cistercian abbey in the valley of the River Rye, west of

Figure 13. The fifteenth-century doorway leading to the Abbots Lodging at Rievaulx Abbey. The 'Annunciation' carving above the doorway was treated with Brethane in 1979 (English Heritage Photo Library).

Figure 14. Detail of the 'Annunciation' carving at Rievaulx Abbey (1997) (English Heritage Photo Library).

Figure 15. The early medieval crosses in the Market Square at Sandbach, treated in 1976 (English Heritage Photo Library).

Figure 16. Close-up of the surface of one of the Sandbach crosses (1997) (English Heritage Photo Library).

Table 11. Scoring table for the Sandbach Crosses (treated with Brethane in 1976, sandstone).

	07/81	04/82	11/84	07/85	10/86	12/87	05/92	06/97	acceptable values
Comparison with untreated areas not possible									
CA	3	1	1	1	1	1	-	1	<3

Helmsley. Analysis of a hand sample has shown the stone at Rievaulx to be an extremely calcareous, fine-grained sandstone, probably from the Middle Calcareous Grit, part of the Coralline Oolite Formation situated near the top of the Middle Oxfordian Stage of the Upper Jurassic. This stone outcrops around the Helmsley area, and has been used locally for building. It consists almost entirely of quartz grains and the main cement is calcite, though much of the cement has been lost. Much of the sample studied in thin section is void space and the stone is extremely porous (overall 25–30%).

The relief carving of the Annunciation, above the fifteenth-century doorway leading into the Abbot's House, was treated with Brethane™ in 1979. Treated stone was described immediately after treatment as having been well-consolidated, but the records made since 1987 suggest a deterioration in its condition. Salt efflorescence was observed in a number of places and appeared to be increasing. Evidence has been noted of voids developing behind some of the carving. This may suggest that the surface has been partially consolidated but that it is becoming detached from the main body of the stone.

Overall, after initial good results, Brethane™-treated areas have begun to absorb water as readily as untreated areas. Compared to untreated areas, the consolidative effect of the system in the first six to eight years was good, but, by the end of the survey, scaling was in evidence and Brethane™-treated areas showed only a slight improvement over control areas.

Sandbach Crosses (Figures 15–16, Table 11)

The monuments comprise the remains of two carved Saxon crosses, incorporating scenes from the life of Christ, and are located in the Market Place, Sandbach, near Crewe, Cheshire. These early medieval crosses were treated with Brethane™ in 1976. The stone is an iron-rich, rather porous silica sandstone of Permo-Triassic age, with large quartz grains. The porosity of stones of a similar geological age and type, from different sites, have been measured at levels ranging from 16–20%, although the porosity of weathered stone may be as high as 25%.

As both crosses were treated in their entirety, comparison with untreated areas was not possible. For the first 16 years after consolidation there was no evidence of weathering and the stone remained in good condition, though with some algal growth in elevated areas. Inspection in 1997 revealed signs of minor scaling beginning to occur on the base of one cross. It was noted also that a slight sheen was present on the surface of both crosses, as well as a generally unnatural appearance. The west side of the smaller cross in particular has exhibited signs of a layer of Brethane™ having been deposited at the surface, possibly caused by the application of too much consolidant.

In general, Brethane™ treatment seems to have had a good, long-lasting consolidative effect, with early signs of degradation only now beginning to appear. A high degree of surface water repellency was evident after treatment, and remained in evidence 20 years later.

Tintern Abbey (Figures 17–18, Table 12)

Tintern Abbey, Gloucestershire, is a twelfth-century Cistercian abbey situated on the River Wye, six kilometres (3 $\frac{2}{3}$ miles) north of Chepstow. The stone is a medium-grained quartzose sandstone ranging from pale

Figure 17. Treated pier on the south side of the nave at Tintern Abbey, showing the treated shaft, second from the left (1992) (Building Research Establishment).

Figure 18. Close-up of the treated shaft, showing present condition of stone and areas where core samples were taken at the time of treatment (1997) (English Heritage Photo Library).

Table 12. Scoring table for Tintern Abbey (treated with Brethane in 1977, sandstone).

	04/80	06/82	09/83	04/84	06/86	02/88	02/92	00/97	acceptable values
AA	2	2	2	2	2	2	2	2	2
AB	3–4	2	2	2	2	3	3	3	<4
AC	0–1	1–2	1–2	1–2	1–2	2	2	2	<2
AD	1	1	1	1	1	2	2	2	<2
BA	1	0	1	1	1	0	1	0	<2
BB	0	0	0	0	0	2	0	2	0
CA	1	2	2	2	2	3	3	-	<3

grey-orange to yellow-brown. It is a medium- to fine-grained lithic greywacke, though approaching a sub-greywacke in composition, consisting mainly of quartz grains in a clay-mineral rich silica cement. The stone is probably from the Old Red Sandstone, possibly the Barbadoes quarry, near Chepstow. The porosity of sample of decayed stone measured in thin section was 10%.

A number of different stone preservative treatments based on acidified silanes were applied to one of the arcade columns of the main nave at Tintern during the 1970s. Brethane™ was applied to a shaft of one of the arcade piers (the second pier from the west wall) on the south side of the nave in 1977.

Monitoring showed the surface to be quite well consolidated at first, but with slight powdering of the surface. Within a few years there was evidence of detachment of the consolidated layer, but the extent of this did not seem to increase significantly in subsequent years. After about ten years reports indicated that the treated stone was generally in the same condition as, and similar in appearance to, untreated stone. In the 20 years since treatment water repellency has progressively decreased.

Conclusions

This study, which has been in progress since 1980, provides an indication of the probable long-term effect of the consolidant Brethane™, and the information gained would seem to confirm that such treatments do have a role to play in reducing rates of deterioration in a range of stone types in exterior conditions. There would seem to be adequate information to confirm that, at least in the first few years following treatment, Brethane™ can effectively reduce biological growth, decay and water

Table 13. Condition and appearance of Brethane treated areas relative to untreated areas after first few years, based on first survey results. X = better ★ = same O = worse

	Years since treatment	Soiling	Biological growth	Decay	Water absorption
Audley End	4	X	X	X	X
Berry Pomeroy Castle	3	X	X	X	X
Bolsover Castle	1	X	X	X	X
Chichester Cathedral	2	★	X	X	X
Goodrich Castle	2	★	X	X	X
Howden Minster	1	★	X	X	X
Kenilworth Castle	1	★	X	★	X
Rievaulx Abbey	1	X	X	X	X
Sandbach Crosses	5	-	-	-	-
Tintern Abbey	3	O	X	X	X

Table 14 Condition of treated stonework relative to untreated areas based on most recent survey.

	Years since treatment	Soiling	Biological growth	Decay	Water absorption
Audley End	20	★	X	X	X
Berry Pomeroy Castle	14	★	X–★	★	X
Bolsover Castle	15	★	★–X	★–X	X
Chichester Cathedral	19	X	X	X	★
Goodrich Castle	19	★	★–O	★–X	X
Howden Minster	13	★	★	★	X
Kenilworth Castle	13	★	X	★	X
Rievaulx Abbey	18	X	X	X	-
Sandbach Crosses	21	-	-	-	-
Tintern Abbey	20	★	★	★	-

absorption. Table 13 shows the results of a comparison of Brethane™-treated and control areas soon after application. It is also important to note that in no case is the condition of the treated area considered to be worse than the untreated and, even more importantly, at least on the ten sites monitored in this programme, that there have been no cases of large-scale failure that may be associated with Brethane™ treatment.

Table 14 shows the results from the most recent survey. The results show that although the treated areas are still benefiting from the use of Brethane™, its effectiveness is declining. This trend can also be seen in the results presented in Tables 2–12, which show that at most sites the biological growth has returned, water absorption has increased and the surface has started to decay after a period of stabilization. The initial decline of biological growth is caused by the toxic effects of the silane monomer. Recolonization may be enhanced in some areas by the presence of a stable surface where Brethane™ was applied. All of these indicate that the Brethane™ in the immediate surface layer (1–2 mm; $\frac{1}{16}$–$\frac{1}{8}$ ins) has weathered, probably by a combination of rain, frost and ultraviolet light and that the surface is now behaving in a similar manner to the untreated control areas.

The results from the exposure trials at the BRE, and elsewhere in field studies (to be reported in a later volume of Research Transactions) tend to confirm that the rate of weathering is still lower than for similar untreated stonework. If this is the case then the outcome of treatment with Brethane™ is that ten to 15 years after application the stonework should be visually indistinguishable from an untreated area, but should be weathering at a reduced rate.

The results also suggest that Brethane™ is more successful when applied to limestones than sandstones. This is unexpected, as at least part of the theory of the impregnation of sandstones with silanes is that they may form mineralogical links with the treated stone. Indeed, further research has identified that MTMOS curing may be substantially retarded in the presence of calcite. The observed success with limestones may be due to the generally higher porosity and higher saturation coefficients of the particular limestones involved. It could also be due to the different decay mechanisms associated with limestones and sandstone, chemical versus mechanical decay. If Brethane™ is more successful at countering chemical decay (for example the effects of acid deposition) then it may be that its water-repellent qualities may be as important as its stone-strengthening properties.

DISCUSSION OF RESULTS

The survey of the sites treated with Brethane™ is important in several ways, as an example of a large-scale, long-term evaluation of a consolidant treatment on architectural elements in varying exposed environments. However, while some important information concerning the performance of one material has been explored and certain trends observed, there has also been much learnt concerning appropriate methodologies for any future survey of this kind.

The most obvious deficiencies stem from the rather basic nature of the recorded information available for each chosen site, combined with a lack of continuity concerning the conservation methodology employed at the sites. While the application pro formas provide a level of quantitative information concerning the treatment, they do not address important qualitative issues that are essential for objective assessment. To a large extent this is due to the fact that the monitoring exercise was based on sites already treated and therefore the base data had not been developed from the project's inception in the detailed way that should be considered essential for such an exercise. The evaluation and relative values of the treated and untreated surfaces are the key to assessment in this kind of exercise and a range of well-considered information types and recording methodologies are required to enable this to take place. During the course of the most recent site observations, it was possible to define some of the insufficiencies in the base data for this particular monitoring programme.

- There was insufficient recorded information concerning the rationale behind the original application of Brethane™ for each site. Not only was the nature of the deteriorated stone surfaces poorly defined, there were no comparisons made between the condition of areas to be treated and those selected as a control.
- The records provided no information on the risk assessment of the Brethane™ application compared to other possible conservation alternatives, and very little on supplementary works carried out as part of the conservation intervention.
- Areas of treatment were poorly defined both in terms of measurement and graphic recording. This has particular significance for comparisons between the treated and the untreated stone, as the boundary layers, being so imprecise, made this extremely difficult.
- The photographic record suffered from a lack of continuity in that there had been no protocol laid down to determine a regime for taking the individual shots in each of the monitoring episodes. Changes in format, position, lighting, scale and colour rendition were unaccounted for as there were no reference markers to register one image to another for comparison.
- In most cases geological identification or analysis of the stones in question was undertaken after, and not at the time of treatment.
- The changes in personnel over the monitoring period meant that there has been an inevitable variation of assessment values based on the criteria as laid out in the tables. This would have not have had such an impact if it were not for the points raised above.
- Over the monitoring period there had been a shift in the perception of Brethane™ as a useful material for the conservation of important stonework. The speed of initial polymerization combined with the exothermic expansion experienced during that process means that other products that avoid these functional characteristics and provide more practical flexibility tended to find more favour. This was rather ironic as the relative speed of initial cure was one of the main perceived plus points during the development of Brethane™. Also during the same period there had been a general backing away from the use of 'synthetic' consolidants, in favour of more 'natural' treatments such as limewater, in the appropriate circumstances.
- On some of the monitored sites the Brethane™ application was not supported by supplementary conservation repair methods and therefore the consolidant was used in a stand-alone role, something which would never be contemplated today. In these circumstances it is difficult to assess the performance of a material which was being asked to perform a broad range of tasks for which it was ill-suited. This unrealistic use of the consolidant detracted greatly from the value of studying the performance of such products on actual sites.

In summary, to obtain the optimum value from such a site-based survey there are a number of prerequisites. The recording of each site and the nature and circumstance of treatment application must follow a clear and consistent methodology, producing comprehensive and objective data in a form which can be translated across a range of stone and design types. While accepting that a site assessment method such as that employed for the Brethane™ trial should, given the foregoing, be capable of providing most of the information required to evaluate treatment performance in general terms, there perhaps should be more safeguards built in to avoid the lack of clarity and subjectivity evident from some of the Brethane™ trial results.

The results of the Brethane™ review have thrown up some surprising conclusions. Firstly, it seems that the limestones treated benefited more than sandstones. This is unexpected as, at least in theory, Brethane™ should be able to form a chemical bond with the silica binding material in a sandstone. That the limestones appear to have benefited more may be due to their greater porosity, allowing a greater degree of impregnation and therefore deposition. This is a little disappointing as it is in the field of sandstone conservation that silanes such as Brethane™ have the most logical application, limestone in the UK today being consolidated almost entirely using lime treatment.

However the next conclusion from the survey may move some way towards breaking down this divide between the two method groups. It appears that while the treated stones weather better than the untreated after several years, the surface of the stone starts to weather back at a similar rate to the untreated. Sampling and experimentation have shown that this deterioration is restricted to the first 1 mm ($^1/_{16}$ in) or so and beyond this the polymer remains intact. This observation may be very important as it opens up the whole area of re-treatment of stones treated with silanes. For if the weathering pattern is only affecting the outer near-surface layers then the maintenance of the stone could be a matter of surface treatment as opposed to the impregnation of the subsurface as in the original application. This surface maintenance may be seen as protecting the deposited polymer as well as the stone and may be carried out with other forms of material up to and including lime-based shelter coats. This type of exercise would be analogous to the use of materials such as Wacker OH to provide a sound substrate on severely deteriorated surfaces prior to the application of, for example, lime-based repair methods.

Such hybrid methods are presently the subject of site use and evaluation on one or two sites in the UK. There is also a case for suggesting that surfaces that have been consolidated with silanes could be coated with a type of compatible shelter coat based on lime products to protect the polymer as well as the stone surface itself. Again, in this way the need to re-treat with a deeply penetrating consolidant may be negated or at least deferred, therefore avoiding the potential longer term problems of material incompatibility and build-up within the near-surface pore structure. There is therefore a strong case for suggesting that future research into this subject should be centred around the use of hybrid systems.

ENDNOTES

1 During the early phase of development of Brethane™ it was recognised that a method was needed to make the silanes miscible with water, in order that an hydrolitic reaction might take place at a reasonable rate. Early experiments involved the addition of 0.01% hydrochloric acid to the water, and agitation of the solution. Hence the term 'acidified' silane (BRE 1976a).
2 An alkoxysilane treatment formerly supplied by Renofors Ltd. This product is no longer available.
3 Supplied by South Western Stone Cleaning and Restoration Ltd. Further information on Wacker OH can be obtained from Wacker Chemicals Ltd, Wacker House, 85 High Street, Egham, Surrey TW20 9HF, UK; Tel: + 44 1784 487800; Fax: + 44 1784 487870.
4 Supplied by Belzona Molecular Ltd. Further information on this product can be obtained from Belzona Polymerics Ltd, Claro Rd, Harrogate, North Yorkshire HG1 4HY, UK; Tel: + 44 1423 567641; Fax: +44 1423 505967.
5 The sole suppliers of Brethane™ (at October 2001) are: Colebrand Ltd, 18–20 Warwick Street, London W1B 5ND; Tel: +44 207 439 9191; Fax: +44 207 734 3358.

BIBLIOGRAPHY

Building Research Establishment, 1976a *Note N73/76, The Use of Alkoxysilanes for the Preservation of Stone. Pt.1. Uncatalysed Polymerization*, Watford, Building Research Establishment.

Building Research Establishment, 1976b *Note N(C)21/76, The Use of Alkoxysilanes for the Preservation of Stone. Pt.2. Catalysed Polymerization*, Watford, Building Research Establishment.

Building Research Establishment, 1978 *Note N(C)18/78, The Assessment of Brethane™ Stone Preservative*, Watford, Building Research Establishment.

Building Research Establishment, 1981a *Current Paper CP1/81, Brethane™ Stone Preservative*, Watford, Building Research Establishment.

Building Research Establishment, 1981b *Note N66/81, Report on the Brethane™ Treatment of the Great West Doorway at York Minster*, Watford, Building Research Establishment.

ANNEX A BRETHANE™ TREATMENT TRIALS: BRETHANE TREATMENT RECORD SHEET (SPECIMEN)

BRETHANE™ TREATMENT RECORD SHEET		REFERENCE (BRE USE ONLY)		
Monument		Berry Pomeroy Castle, Devon	Date of erection	
Exact location of treatment		pier of loggia (east end)		
General description of subject		lower remaining part of loggia pier on granite base		
Aspect		south		
Type(s) of stone		limestone, beer	Jointing and pointing	
Dimensions of area to be treated		2'0" (h) x 3'7" (l) x 1'10" (d): segment 6" (l) x 1'6" (d) removed from centre length portion. Decorated face south		
Person in charge		B P Hodgman DAMHB		
Operative(s)		R Jones, B Trewitt, G Wilmott, W Dyke, all DEL & DAMHB		
Details of any surface preparation, eg cleaning, carried out before Brethane treatment		light brushing with bristle brushes		
Details of measures taken to ensure thorough drying of stone prior to treatment		scaffolded and sheeted canopy erected over stone in November 1982		
Date(s) of treatment		28/6/83		
Weather on day(s) of treatment	Temperature	75 degrees F		
	Wind direction	SW		
	Wind strength	weak		

Details of treatment
  Formulation used
  Method of application
  Volume of Brethane applied
  Area treated
  Time taken
  Curing time of sample(s) of Brethane
		standard Brethane™
		spray
		7.6 litres
		22 square feet (2.04 square metres)
		3.25 hours (0945 to 1300)
		approx 4 hours		

General observations at time of treatment. Please give details of any suspected rising damp or soluble salt contamination		suspected rising damp to stone and adjacent corework is assumed cause of low Brethane intake. We hope to take core samples in Sept/Oct 83 to determine penetration depth		
Details of any weather protection given to the treated area after treatment and when this was removed		canopy to be struck on 12/7/83		

Building Research Establishment, 1983 *Note N110/83, Brethane™: Packaging and Shelf Life*, Watford, Building Research Establishment.

Butlin R N, Yates T J S, Ridal J P and Bigland D, 1991 Studies of the use of preservative treatments on historic buildings, in *Science Technology and European Cultural Heritage. Proceedings of the European Symposium, Bologna, 13–16 June 1989*, (eds) Baer N S, Sabbioni C and Sors A I, Oxford, Butterworth Heinemann Ltd, 664–7.

Butlin R N, Yates T J S and Martin W, 1995 Comparison of traditional and modern treatments for conserving stone, in *Preprints of the International Colloquium on Methods of Evaluating Products for the Conservation of Porous Building Materials in Monuments, Rome, 19–21 June 1995*, Rome, ICCROM, 111–120.

Clarke B L and Ashurst J, 1972 *Stone Preservation Experiments*, Watford, Building Research Establishment/Department of the Environment.

Ridal J P, 1988 *A Determination of the Penetration of Two Consolidants and a Water Repellent into Stone Cores, removed from English Heritage Buildings, using Infra Red Absorption Spectroscopy*, Building Research Establishment Report N 128/88, Watford, Building Research Establishment.

ACKNOWLEDGEMENTS

The monitoring that forms the experimental part of this research has taken place over the last 20 years and consequently many people have been involved in it. The authors would like to acknowledge the help of Keith Ross, David Bigland, Julian Ridal, John Houston (all of BRE), John Ashurst, Brian Hodgeman (both formerly of RTAS, EH) and the many staff at the sites involved.

AUTHOR BIOGRAPHIES

Bill Martin trained as a stone conservator and ran his own studio for nine years before moving on to the Council for the Care of Churches where he was Conservation Officer. He joined English Heritage in 1989 and his responsibilities include project management of research into stone consolidants, the decay and conservation of historic tile pavements, as well as coordination of technical advisory work for the Building Conservation & Research Team. He managed the English Heritage Metals Conservation Studio in Regents Park prior to its closure.

David Mason trained in fine art, has worked as a stone conservator, and was a Research Fellow at De Montfort University, Leicester, specializing in the history and theory of conservation. He worked in Italy, gained a PhD in 1997, and in the same year joined what is now the Building Conservation and Research Team at English Heritage as an Architectural Conservator. He has co-ordinated the BCRT research programme, and was series editor of the Research Transactions between 1999 and 2001.

Jeanne Marie Teutonico was formerly on the staff of the International Centre for the Study of the Preservation and the Restoration of Cultural Property (ICCROM), in Rome. She became Senior Architectural Conservator at English Heritage and managed the penultimate monitoring and evaluation phase

ANNEX B. BRETHANE™ TREATMENT TRIALS: CONDITION REPORT SHEET (SPECIMEN)

Site	Chichester Cathedral	Reference	
Location of trial	Central tower, lantern area, east side, second opening from north. Quoin stone		
Date of treatment 3/78	Date of inspection 6/6/86	Inspected by B Hodgman	

Refs of photographs taken BPH/A/-1

APPEARANCE OF TREATED STONEWORK	✓ as appropriate	remarks
lighter than untreated stonework		
much the same as untreated stonework	✓	
darker than untreated stonework		
dirtier than untreated stonework; streaked		
uniformly dirtier than untreated stonework		
much the same as than untreated stonework	✓	
uniformly cleaner than untreated stonework		
cleaner than untreated stonework		
no biological growth	✓	
less biological growth than untreated stonework		
same biological growth than untreated stonework		
more biological growth than untreated stonework		
generally less decayed than untreated stonework	✓	
much the same decay as untreated stonework		
generally more decayed than untreated stonework		

CONDITION OF STONEWORK	treated stonework	untreated stonework	remarks
no powder transferred to finger drawn lightly across surface			very friable and spalling appearance but quite firm to touch. No evidence of loose or spalling flakes
slight amount of powder transferred to finger drawn lightly across surface	✓	✓	
substantial amount of powder transferred to finger drawn lightly across surface			
traces of stone fall away when finger is drawn lightly across surface			
surface scales present, less than thumb nail size			
surface scales present, greater than thumb nail size			
surface cracking and crazing			

WATER REPELLENCY OF TREATED STONEWORK	✓ as appropriate	remarks
absorbs water as readily as untreated stonework		
absorbs water less readily than untreated stonework		
does not absorb water: film of water remains on surface		
does not absorb water: water sits on surface in globules or runs straight off	✓	

OTHER OBSERVATIONS	Well consolidated and no change from previous inspection on 22/6/84

of project AC2 on masonry consolidants, and contributed to the interpretation of final data. In 1999 she joined the Getty Conservation Institute and is now its Associate Director, responsible for field projects and research.

Sasha Chapman trained as an archaeologist specialising in recording buildings. She joined English Heritage in 1993 as an architectural conservator and was responsible for coordinating technical enquiries relating to graffiti removal from historic buildings and monuments. She is a former Chair of the United Kingdom Institute for Conservation of Historic and Artistic Works, Stone and Wall Paintings Section.

Roy Butlin is a physical chemist with a background in industry, academia and the scientific civil service. After ten years at the Fire Research Station at BRE, he spent two years at the headquarters of the Department of the Environment, then rejoined BRE, where he set up the Weathering Science section in 1984 to study the effects of acid deposition on building materials. From there his interests expanded into a wider programme of work on heritage matters, including the launch of the Heritage Support Service at BRE, of which he was manager. He retired in 1997.

Tim Yates is Director of the Centre for Heritage, Stone and Masonry Structures at the Building Research Establishment Ltd. He spent four years as a field archaeologist before reading Archaeological Sciences at Bradford University and then spending three years at University College London researching the analysis of Holocene carbonate sediments. He joined BRE in 1986 to compile and edit a report on the effect of acid deposition on buildings and building materials and now leads a team researching heritage buildings, stone, masonry and mortar.

Lime treatments
An overview of lime watering and shelter coating of friable historic limestone masonry

JOHN FIDLER
English Heritage, 23 Savile Row, London W1S 2ET, UK. Tel: +44 207 973 3025;
Fax: +44 207 973 3130; email: John.Fidler@english-heritage.org.uk

Abstract

Lime treatments, used individually or in concert throughout history, have been the focus of renewed interest at the turn of the millennium. Methods, trends and previous evaluations are described. Current and future research will determine the treatments' contribution to the preservation of decaying historic limestone masonry and sculpture.

Key words

Lime treatment, lime watering, lime shelter coating, lime method of masonry consolidation

INTRODUCTION

Lime water and shelter coats are two components of a more extensive lime-treatment system commonly used for the surface repair, consolidation and maintenance of historic limestone[1] buildings in England. Other parts of the system, used individually or as sequential processes, include lime mortar filleting and grouting, and lime-mortar dentistry patching of decayed stonework. Another component, the hot-lime poultice once adopted for cleaning purposes, is now rarely used. All the techniques have been employed with varying degrees of perceived success for many years but have never been scientifically proven (Price 1984, 160–62; Price & Ross 1984, 301–312; Price et al 1988, 178–186; Price & Ross 1990, 176–184). This paper reviews past, current and future developments.

DEFINITIONS

Lime watering involves the much-repeated application of very dilute calcium hydroxide in solution to friable masonry, with the objective of consolidating the material through carbon dioxide absorption and re-conversion to calcium carbonate, the primary cementing matrix of most United Kingdom limestones (Fig 1).

Shelter coats, designed as buffers or sacrificial layers, are often applied as the final treatment in the repair process, and consist of thin applications of a mix of dilute, freshly-slaked non-hydraulic lime and finely crushed aggregate, sometimes supplemented by modest additions of casein and formalin.

Lime washes are applied decorative coatings of thinner (more dilute) consistency than shelter coats and thicker (less dilute) consistency than lime water, often incorporating pigments and sometimes tallow in traditional recipes. They are not discussed here.

Figure 1. Stone consolidation. The west front of Wells Cathedral, in Somerset, UK. Stone statuary here has been subject to two forms of masonry consolidation: in the case of the east elevation of the north-west tower, treatment with an alkoxysilane, and for the majority of the West Front, various kinds of lime treatment. The sites are regularly monitored as part of the local site maintenance and as part of the national consolidant trials run by English Heritage (Photograph by Jerry Sampson).

HISTORICAL DEVELOPMENT

Lime wash coatings have been used as a form of masonry protection and decoration in Britain since Roman times. At Caerleon in South Wales, for example, a Roman legion used lime wash as protection for its hastily erected and poorly constructed masonry camp walls. The Roman builders also took advantage of this surface to deceive the Welsh tribes by painting the walls with red ochre lines to suggest the dressed masonry jointing of a more robust line of defence. The famous White Tower at the Tower of London, built in the eleventh and twelfth centuries, was aptly named for its method of maintenance: the periodic external repainting of the rubble stonework with lime wash (Salzman 1952, 157). King John later decreed that lime washing should be used extensively across London, as a means of repair and (it is said) as a way of fireproofing timber-framed construction. Many medieval masonry buildings were scraped in the eighteenth and nineteenth centuries so there is little evidence of how extensive this maintenance process had become. Yet craft traditions persist, and at Westminster Abbey generations of masons have continued the guild practice of lime shelter-coating the stonework in most campaigns of repair and replacement.[2]

In the Victorian era, especially following the debacle over the choice of inferior stone for the rebuilding of the Palace of Westminster, numerous patents were taken out for stone consolidants, some involving earth hydroxides. For example, A H Church patented a recipe using barium hydrate for limestone and marble preservation in 1862.[3] Reigate stone (a calcareous sandstone) was treated with calcium caseinate by Sir George Gilbert Scott on the interior of Rochester cathedral.[4] At the turn of the twentieth century, Arts and Crafts architects, such as W R Lethaby, and proponents of the Society for the Protection of Ancient Buildings (Powys 1929, 95–99) rediscovered and specified those ancient traditions and advocated their revival for conservative building preservation. Experiments were performed with adjacent natural coating systems, and Lethaby, for example, suggested the use of distemper (crushed chalk whiting, water and animal glue size with added pigment) for stone conservation at Wells Cathedral in Somerset.

More recent interest in the development of lime-treatment systems stems from the research of the late Professor Robert Baker and his trials at Bristol. These were on the facade of Marshall Wade's House in Bath, and at Merton College, Oxford from the mid 1950s onward, until the culmination of his efforts in the comprehensive conservation plan for the thirteenth- and fourteenth-century statuary on the West Front of Wells Cathedral during the late 1970s and early 1980s (Caroe 1985, Caroe 1987).

In the summer of 1950, decaying Doulting stone on the north porch of Wells was cleaned by water washing and a first attempt was made to employ lime watering. The experiment was extended to the West Front of the cathedral in May 1952 when the north aisle door was similarly treated. Later, in 1959, figure sculpture (statues S176 and S236) was subjected to the same regime. On 9th October 1965 a meeting was held at Wells between Alban Caroe, Surveyor of the Fabric (1974–1), Bert Wheeler, Cathedral Clerk of Works (1974–8) and Professor Baker (later the consultant conservator 1977–81) to decide on experiments using lime water which could be monitored by the Building Research Establishment. Trials with lime poulticing and lime watering by Baker were then undertaken on statues S383 and S283 between August and November 1968 (Sampson 1998, 3 and 4). Subsequent monitoring over the next decade revealed that the surfaces of the treated figures remained 95% sound, whereas the adjacent controls which had been simply washed continued to decay fast.

Concurrent with, and in contrast to, Baker's work at Wells, limited trials were undertaken there with alkoxysilane masonry consolidants, then considered by the West Front Conservation Committee to be experimental and risky, due mostly to their irreversibility. In 1974–5, Ken Hempel of the Victoria and Albert Museum, London, applied Rhone-Poulenc X54-802 to statues S32 and S177 on the eastern elevation of the north-west tower. In 1977, John Ashurst of the Directorate of Ancient Monuments and Historic Buildings, Department of the Environment, and Clifford Price of the Building Research Establishment deployed Brethane on sculpture S117. Finally, a demi-angel was taken into museum conditions and treated by the cathedral's own conservators with an acrylic/silane consolidant based on Dow Corning products, on the advice of John Larson of the Victoria and Albert Museum (Sampson 1998, 271). While these trials have stood the test of time without detriment to the statuary,[5] the Committee at the time recommended the adoption of the Baker method for the majority of the work because it deemed fewer risks to be involved.

From the sculpture conservation atelier of this large-scale, pioneering campaign, the lime treatment method has spread far and wide. Its advocates praise its appropriateness in terms of historical precedent, craft tradition, natural process and technical efficacy, replacing like-for-like in the decaying matrices of the stone (Figs 2, 3 and 4).

As its popularity has grown, so too have the myths surrounding the process and its claimed effectiveness. Yet scientific scrutiny has so far failed to explain fully the phenomena and its patently demonstrable superficial benefits. This paper reports on past and current practice and describes recent and intended future technical evaluations.

FACTORS FOR SUCCESS

R J Schaffer (1932 [1972], 36) described the factors critical to the success of stone consolidants more than sixty years ago. The material should increase the stone's tensile and shear strength and reduce its porosity and absorption, while permitting the transfer of internal moisture out and should harden the stone surface and immediate substrate, making it resistant to salt crystallization. Ideally, the treatment should consist of material whose composition is identical or very close to that of the stone being conserved and should not be visible or

Figure 2. Wells Cathedral, West Front decay: a fifteenth-century statue of an angel from the gable. The late Professor Baker and architect Martin Caroe faced the problem of gypsum crusts and cavities caused by sulphate damage to limestone statuary (Photograph by Jerry Sampson).

Figure 4. The newly carved copy of sculpture No. 160 by Derek Carr made of Doulting stone and installed in 1984. The original of the medieval king fell in the 1850s and had been put back together with over one hundred iron dowels that rusted, expanded and caused the stone to fall apart (Photograph by John Fidler, English Heritage).

Figure 3. Alternative treatments: alkoxysilane trials at Wells cathedral. The two left-hand sculptures have been consolidated. Statue S117 (St Theopistis) was treated with Brethane™ in 1977 by DAMHB/DOE and BRE; Statue S177 (seated Roman Emperor) below was treated with Rhone Poulenc X54-802 in 1974–5 by the V&A Museum. (Photograph by John Fidler, English Heritage).

change the colour, texture or surface reflectance of the historic masonry.

Advocates of the lime treatment suggest that the various elements of the system fulfil these criteria. Lime watering can tighten or harden friable surfaces and, since it consists of material similar to that within stone, it should decay in the same way as the subject masonry. The lime treatment is not claimed to reduce porosity: the reverse is said to take place. The stone is permitted to 'breathe' i.e. damaging salts are allowed to effloresce at the surface, rather than within the upper pore structure.

Shelter coating is not claimed to be a consolidant. It conceals the original surface under a coating and thereby protects it physically and chemically from immediate attack from the elements. It provides a sacrificial barrier and can be maintained by rubbing down and re-coating periodically.

LIME TREATMENT METHODS

Hot lime poulticing

In the original Baker lime method, a hot lime poultice was applied to the decayed and gypsum-crust-covered limestone to aid the cleaning and removal of damaging salts and dirt and to open the pores of the stone's surface for subsequent treatment. Although the technique received a great deal of public and professional interest at

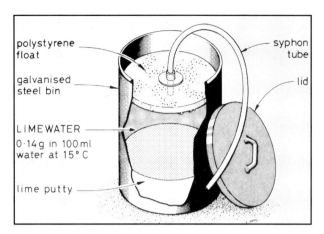

Figure 5. Slaking lime (drawing by Iain Mc Caig, English Heritage).

Figure 6. A lime slaking facility on-site in the UK. Although lime putty for mortars can now be acquired direct from manufacturers and wholesalers in tubs, there are many groups that favour processing the materials from quicklime. Here we see the slaking trough at the highest point, where calcium oxide is added to water, and, below it, various settling tanks that enable lime water to be run off for use separate from the putty (Photograph by Iain McCaig, English Heritage).

the time, it was abandoned by the West Front conservation team before the large-scale works were put in hand and probably only four figures were actually so treated.[6]

Lime watering

After cleaning [7] friable limestone is often treated with lime water, which many conservators perceive to be a natural consolidant. In many cases it definitely appears to strengthen the surface, and powdering, soft decayed material feels toughened. In some situations, over-application has led to the production of a strangely glassy surface.

How is lime water manufactured? When quicklime is added to a shallow tank of water and slaking takes place, the resultant products have varied uses. Over-burnt limestone is thrown away. The lime putty is raked and sieved off via settling tanks to drain and be used for mortar binders and, in dilute form, for shelter coats. The milky residue is tapped as lime water. A simpler technique (Fig 5) is to pour lime putty into a container of water and shake it up, leaving it to settle under a sealed lid or polystyrene float to avoid carbonation (Ashurst & Ashurst 1988, 82).

Lime water contains very small quantities of calcium hydroxide.[8] When lime water is applied to the stone, it is exposed to carbon dioxide and changes back to calcium carbonate. The minute particles in solution expand in volume and weight during the absorption of the gas. Theory has it that a tight fit is made in the pore structure for the deposition of material similar to the stone's own cementing matrix.[9] Application is achieved by sieving off the clear solution over the lime putty in the clearing water into spray bottles for spraying onto the friable stone. Many applications are needed, and up to forty seem to be the norm undertaken for several days. Application can continue as long as the surface will absorb, but excess lime water should not be allowed to lie on the surface of the stone. It can be removed by sponges squeezed out in clean water. Obviously, the drier the stone before treatment, the more absorption is likely to take place. Success has been reported mostly on oolitic limestones, but chalk and magnesium limestone and lime plasters have also been treated.

Attempts in Britain to scientifically record or quantify the effects of lime watering have so far failed (Price et al 1988, 178–186; Ashurst 1990, 182). Tests carried out on Doulting stone from the cloisters (c 1470) at Wells Cathedral were perhaps not sensitive enough to record subtle changes experienced by the stone. No calcium hydroxide or new calcium carbonate deposits could be seen using petrological thin section examinations. Researchers suggest that the very thin layers could have already converted to calcium sulphate, which itself would be indistinguishable from pre-existing sulphate deposits. In fact, early research showed that a 'consolidating' effect similar to that produced by lime water could be produced by the multiple application of distilled water (Clarke & Ashurst 1972).

It is further suggested that the surface consolidation might have as much to do with the solution and even redeposition of calcium sulphate as it does to the introduction of lime water. This seems to be borne out by the recent experience in England of conservator Paul Harrison, who verbally reported to English Heritage his perceived comparative success in 'consolidating' magnesian limestone. As magnesian sulphate is more soluble than calcium sulphate, it can be more readily transported to, or created close to or on, the surface, resulting in surface hardening.

Efforts to measure increases in strength or resistance to abrasion in lime water-treated material have not yet succeeded (Price & Ross 1990, 182). The results were inconclusive because of the non-homogeneity in the subject stone (there was a lack of uniform original decay and inconsistencies in the applied treatments) and because of statistically inadequate numbers of samples. No discernable effect could be seen in changes of overall porosity. The consolidating action was inconclusive or 'not significant', despite radioactive tracer plotting of deposition up to 26 mm (1 in) within the structure and more than 50% of the lime being deposited in the outer 2–3 mm ($1/_{16}$ in) near the surface (Ashurst 1990, 183).

Micro-repairs in lime mortars and grouts

Following 'consolidation', mortar repairs are placed to make good damaged areas, to grout blisters and voids and

Figure 7. Basic dentistry repairs. A conservator working for the contractor Rattee and Kett, of Cambridge, uses lime-based dentistry mortar repairs for infilling stonework decay pockets on the exterior of Henry VII's Chapel at Westminster Abbey prior to shelter coating, as part of the repair works specified by the aptly-named Surveyor of the Fabric, Donald Buttress, FRIBA (Photograph by John Fidler, English Heritage).

Figure 8. Limestone decay. Sulphates and ruptured gypsum crusts are in the foreground with cavernous powdering decay behind (Photograph by Jerry Sampson). See Colour Plate 5.

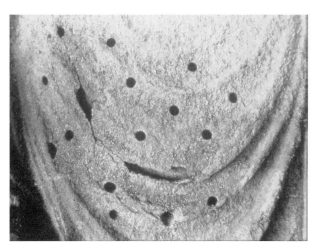

Figure 9. The lime technique. Holes are drilled for access for flushing out harmful salts and for the introduction of grout and mortar dentistry repairs, as a precursor to lime watering and shelter coating. Only a comprehensive approach to the lime technique, with ongoing maintenance, will succeed. Sometimes the treatment is not as gentle as is claimed (Photograph by Jerry Sampson). See Colour Plate 6.

to re-attach lifting or flaking crusts. At Wells, all repairs were based upon mature, high calcium, non-hydraulic lime putty binder without cement. If a weak hydraulic set was required, then powdered brick dust or HTI powder [10] were used as a pozzolan. The properties of these materials are now better understood (Teutonico et al 1994) and a range of indigenous and imported hydraulic limes and trass have now been added to the conservators' pallet (Teutonico 1997).

In addition to the binder and an occasional setting agent, aggregates are selected and graded for colour, texture and function. Sands and crushed stone are carefully blended, and tests are made to find the right match for the weathered subject stone. As an adhesive, the mortar blend is normally in the ratio 1:1 by volume, lime to aggregate. Grouting deploys a weaker mix of 1:1.5, and repair mortars are weaker still at 1:2. Of course, the particle size and size distribution of the aggregates vary according to the function of the mix; for example, smaller for the grouts and larger and more well-graded for the mortars (Ashurst & Ashurst 1988, 83–84).

Upwards of 32 or more mortars might make up a conservator's palette before starting work on the repair of delicate, valuable masonry in this fashion. Cavities and cracks are flushed out with hand sprays and the surfaces moistened to avoid suction on the repair. Deep cavities are treated with a slurry of the repair mortar, followed by a filling of mortar and small pieces of limestone to limit the depth of the repair. Layers no more than 6 mm ($^1/_4$ in) deep of the patch repair are pushed into position using dental tools and spatulas and pressed home, as good compaction of the amalgam is essential, just as it is in dentistry. The repair is then covered to protect it from rapid drying by direct sunlight or wind, and the area is kept moist by periodic spray wetting. While the material is still green, a variety of tools, sponges and rags are used to finish the surface of the overfilled cavity.

Protection from the weather is essential if the weak, non-hydraulic mortar is to set by carbonation alone. Grouts need pozzolanic assistance as they are buffered from the air deep within fissures and blisters. In many cases, new punctures are needed in fissured and blistered stonework to gain access to rinse out gypsum, eliminate

Figure 10. Lime watering. Many applications are necessary before the calcium hydroxide in solution appears to have any effect, though scientific testing has still to prove that it actually works. It is possible to get the same surface hardening by using tap water (Photograph by Bill Martin, English Heritage). See Colour Plate 7.

air pockets, and gain keys for the repairs (Fig 9). They can disrupt the appearance of the stone and can be concealed at a later stage only by expert cosmetic patching. In all cases, aftercare and maintenance are essential prerequisites for a successful repair.

Shelter coating

The final stage of the lime treatment is the application of a thin surface coating to the cleaned, consolidated and repaired material to act as a physical barrier, chemical buffer and sacrificial layer. The shelter coat is of similar or identical composition as the repair mortars. The lime to aggregate proportion is slightly higher, the aggregates are finer and water is added to make the consistency of thin cream. A typical mix would be 3:8 lime to aggregates, the latter falling through a 300 µm sieve.

Some traditional recipes have included skimmed milk (casein) in the mix (by volume 1–3:10, skimmed milk to dilute lime putty), as this mixture produces calcium caseinate, which has coagulating, adhesive and water-repellent qualities. The casein can provide a food source for biological activity, so formalin is added to help sterilize the mix. [11]

Other shelter coat recipes seem to perform perfectly well without the additives. The subject stone is sprayed with water until no suction is seen. Then the shelter coat is laid on with a soft bristle brush and worked into the texture of the stone. Individual stones are treated by themselves, not with a painting technique applied across the mortar joints, which would affect the overall character and appearance of the masonry.

The coating, and lime watering, are not recommended for use on most sandstones, as the breakdown of the surface caused by exposure to sulphur dioxide produces calcium sulphate, which is highly damaging to most of their ferruginous and argillaceous cementing matrices (Fig 10).[12]

The coating fills small crevices and depressions. Drying out is carefully controlled, as rapid drying can lead to a powdering surface and undesirable colours. If too much coating is applied, arises and decoration can be obscured, and the colours and textures produced can look overly artificial and smothering to the natural finishes and patina of the stone. Inexpert application can lead to a bland and woolly appearance. Properly expedited, however, the thin layer can enhance the character of the original surface.

CURRENT SITE PRACTICE

Although originally conceived by the revivalists, especially at Wells Cathedral, as a comprehensive system, the lime treatment method is now becoming compromised as conservators pick and choose from the range of techniques to suit their needs and perceptions. Few now practice poulticing, especially with slaking lime, but lime watering continues to grow in popularity, as some professionals become disillusioned with polymers because of their relative costs, limitations of use and the onerous health and safety requirements necessary for their application. In some quarters, there is also something very appealing in the use of a traditional process that lacks sophistication. These people believe in the process and in the results they can see and feel, not in what science can prove or disprove.

Lime watering is now commonly assumed to be the only part of the system worth implementing. Yet in good practice, it is rarely undertaken without supplementary surface repairs and must be supported by regular maintenance. And importantly, the performance of lime watering should not be judged in isolation from the comprehensive treatment and its aftercare. In fact, there are many conservators who use filleting, grouting and mortar patch repairs, and shelter coats but prefer not to use lime-water consolidation.

John Ashurst (Ashurst 1990, 174) has likened the unfailing belief in lime water to 'The Emperor's New Clothes'. But it behoves those responsible for setting national technical policy, such as English Heritage, to properly understand the phenomenon and to give clear objective advice.

RECENT, CURRENT AND FUTURE RESEARCH

In 1994 English Heritage commissioned research from the Department, now School, of Conservation Sciences at Bournemouth University to address some of the site-related questions concerning the effectiveness of the lime treatment system. Project AC9, called 'Limes and Lime Treatments', complemented other research into the decay processes of various English building stones, studies of the performance of lime-based mortars and of the effectiveness of certain alkoxysilane consolidants.[13]

The first stage involved the comprehensive mapping of all the sources of currently available building limes in the British Isles, whether they were manufactured or imported, and to collect data or expedite testing on their characterization and performance. Specifiers, contractors and conservators continue to write and speak about 'lime' as though it were one standard material, but there are many different varieties and ways to describe their properties, depending on whether vernacular craft, professional/technical or scientific terminology and understanding are being used.

From this work, English Heritage has published a national directory of sources of supply (Teutonico 1997) with the first-ever direct comparison of orders of cost. It has also been contributing to the European development of scientific classifications for building limes, the first comprehensive review since 1929. The work has culminated in the long-awaited publication by the British Standards Institution of BS EN 459:2000 on *Building Limes* (British Standard 200a).

In the second phase of work, prominent long-standing lime treatment sites were revisited. The original conservators or masons who performed the work were asked to compare, by questionnaire and detailed close-range site inspection, whether the process continued to perform relative to nearby untreated sites. Obviously, the subsequent maintenance of treated areas had to be taken into account, and the overall effectiveness of the study is only as good as the original pre- and post-treatment site records and the conservators' (subjective) memories.

Nonetheless the study (Woolfitt & Durnan 1995, see also Woolfitt, this volume) does log expert conservators' perceptions of their actions: an important factor when trying to understand developments in past and current practice and their cost/benefits.

As part of the work collecting data for English Heritage's National Register of Treated Sites, they have also been planning to establish objective lime-treatment trials at the organisation's own sites for posterity's benefit and information. Detailed assessments of the subject materials, decay processes and treatment regimes before, during and after treatment need to be recorded and they have been developing a decay mapping and monitoring protocol, using stereo rectified photographic imaging and computer-aided drafting.

Currently English Heritage is also reconsidering Clifford Price's scientific tests for evaluating the effectiveness of lime watering (Price & Ross 1990, 176–184) with the hope of devising much more precise and sensitive tests with statistically relevant sampling for possible new laboratory work in the immediate years ahead.

For example, Price's team employed a technique devised by Butterbaugh (Price & Ross 1984, 307) to

Figure 11. Lime shelter coat. The long-term maintenance of the shelter coat is a prerequisite of good practice (Photograph by Bill Martin, English Heritage). See Colour Plate 8.

measure the abrasion resistance of treated stone analogous to strength testing. But the force exerted on the lime-treated surface in the tests was 2.75 bar (40 psi), too strong to show any conclusive difference in results. Perhaps much lower pressures would be able to highlight expected differences? Also, the abrasion holes created by the test had to be accurately measured for changes and 120 mesh carborundum grit was employed to fill the cavities and weighed to establish the volume. Now new laser surface profilometers should be able to record surface changes more precisely.

It has also recently been pointed out (Brajer & Kalsbeek 2000, 145–156) that the crystal structure of calcium hydroxide changes as it goes in and out of solution during the forty plus passes of the application process and makes further time-dependent changes subsequently during the carbonation process. Perhaps previous attempts to prove/disprove the effectiveness of lime watering by laboratory testing took place too soon after initial application? Certainly the speed and efficiency of carbonation may have significant effects and English Heritage has noted recent experiments in sculpture consolidation by John Larson at the National Museums and Galleries on Merseyside in this direction.

Obviously, current and future research must pay attention to trends in site practice. Already we are beginning to observe hybrid treatments developing which employ a combination of lime- and polymer-based techniques and these must be taken into account when assessing the benefits or disbenefits of so-called lime treatment regimes.

So as to gain international consensus on the key research questions to be answered in the next five years, a scientific workshop for European and American specialists in the field was held in London in December 2000, organised by English Heritage and the Getty Conservation Institute. The two-day event focused on defining the boundaries of interest, and current state-of-the-art, in masonry consolidation that employs earth hydroxides, tartrate conversion layers and bio-activation of calcite crystal growth. Steps are now in hand to summarize the proceedings and submit them for possible publication in the International Institute for Conservation's *Reviews in Conservation* series.

CONCLUSION

Lime treatments, including lime watering and lime-shelter coats, are partial means to extend the life of decaying limestone masonry and sculpture. Used throughout history and revived in the second half of the twentieth century as part of a broader, sensitive approach to the care of monuments (particularly connected with the conservation of medieval statuary on the West Front of Wells Cathedral in England), the processes patently have some effect that cannot yet be scientifically modelled or quantified.

Despite the best efforts of leading authorities in the field, no conclusive evidence exists to prove whether lime watering works as a surface consolidant. Clearly, there is some beneficial change to decaying surfaces from its use, but this can also be replicated by the application of distilled or plain tap water.

Further research is undoubtedly required to understand the phenomena. English Heritage, together with its research collaborators and contractors, continues to work to settle one of the most puzzling outstanding questions for conservation today.

ENDNOTES

1. That is, limestones formed principally of calcium carbonate: lime treatments are not recommended for use on calcareous or other sandstone types as they can contribute to their decay.
2. For example, the last Surveyor of the Fabric, Donald Buttress FRIBA, employed shelter coats during his restoration of the West Front of the Abbey in the 1990s in order to preserve Henry Yevele's fourteenth-century external porch vault between Hawksmoor's and James's eighteenth-century Portland stone west towers. Shelter coats were also employed by him for the repairs to the exterior stonework of Henry VII's Chapel at the east end.
3. Otherwise known as Baryta Water or barium hydroxide, the material is said to eliminate damaging gypsum and forms an insoluble barrier. Calcium sulphate combines with barium hydroxide to form insoluble barium sulphate and calcium hydroxide, the latter then carbonating to become calcium carbonate (Church 1862). The treatment was popular up until the turn of this century (Church 1904) and has been revived since, but there have always been risks of crust formation without depth of penetration, and consequent loss of fabric (Martin 1996).
4. Information supplied in personal correspondence from the late Martin Caroe RIBA, Surveyor to the Fabric of Rochester Cathedral, 9th November 1998.
5. According to Caroe, who was the architect for the Wells West Front programme (pers comm, 9th November 1998), all the external alkoxysilane treatment trials were showing some loosening of grains or very local breakdown of the surface after twenty years of exposure and weathering. This evaluation, based on regular cycles of visual inspection and comparisons of photographs, coincides with English Heritage's own findings from 18 years of monitoring Brethane-treated sites throughout England (Martin et al, this volume).
6. For the record, fresh quicklime (calcium oxide) was broken into pieces to pass a 12 mm ($\frac{1}{2}$ inch) sieve and was bound against the subject stonework with sheets or strip sacking in a layer 12–25 mm ($\frac{1}{2}$–1 ins) thick. The edges were then sealed and the ensemble covered in polythene. Then water was poured in at the top so that the slaking took place. Slaking is the term used to describe the addition of quicklime to water: calcium oxide plus water produces heat, calcium hydroxide and carbon dioxide with resultant expansion in volume. Extreme heat is given off through this chemical process and the materials should only be mixed, the lime to the water, a bit at a time, in a galvanized steel or other robust container. Protective clothing including gloves and goggles are essential to prevent alkaline burning, and it is highly dangerous to reverse the order and add water to a pile of quicklime (Teutonico 1997).

   The medium was then left on the stone for seven to ten days and kept wet before being carefully cleaned off by spraying with water. The process was dangerous for operatives and risky for the monument since excessive heat could cause the hessian to catch fire, and swelling could lead to the bursting of the poultice pack. Now, if it is used at all, this process has been modified so that the fresh slaked lime is applied as hot paste on a pre-wetted surface by hand using a trowel. The poultice is left on the stone for two to three weeks, and the plastic sheet is lifted periodically and the paste and hessian wetted continually to keep the poultice working.

   Dirt is raised and softened on the surface and then taken off with intermittent nebulous water sprays and/or with air abrasive pencils. Critics say that the poultice and water-washing system can endanger overlying painted gesso decoration, if it is not first identified and protected. Conservators reported that there was less risk of damage and loss to the polychromy on the West Front of Wells Cathedral if air abrasive pencils were used instead in sensitive locations. There seems to be very little difference in the efficacy of the poultice concept between standard attapulgite clay media or active ammonium carbonate and the lime method.

   Commentators suggest that the poultice and washing do not extract calcium sulphate from the depth of the stone, as is sometimes claimed. However, these methods can remove surface skins of gypsum and do open up the surface pore structure to create a more absorbent material (Price & Ross 1990, 182). This can be damaging for the stone if no further remedial treatment is applied. The open pores increase the material's surface area which is then susceptible to increased wet and dry deposition of pollutants, potentially resulting in the enhanced re-formation of gypsum. The hot-lime poultice has absolutely no consolidating effect. It is only one of many cleaning techniques that can be deployed and the alternatives have much fewer conservation and health and safety risks associated with their use.

7   For masonry cleaning techniques in general see British Standard 8221 & 2: 2000b. At Wells Cathedral, nebulous mist water sprays were applied to robust areas of plain stonework to soften the gypsum crusts and soiling for ease of removal by hand bristle brushing and low-pressure, low-volume rinsing. However, on highly friable areas of sculpture where there was evidence of gesso and decorative painting, the conservators preferred to use micro-abrasive 'pencil' techniques to clean the stone.
8   Ashurst & Ashurst 1988, 81–82. The authors suggest only 0.14 g in 100 ml of water at 5 °C.
9   When exposed to CO_2 the $Ca(OH)_2$ converts to $CaCO_3$. In a litre of slaked lime solution holding 1.7 kg (4 lbs) of solid, the conversion produces 2.3 kg (5 lbs) of calcium carbonate. Most British building limestones have a porosity of 20% for material with a solid density of 2720 kg/m^3 (7 lbs/ft^3) So the deposition of $CaCO_3$ by lime water at the above-mentioned rate would mean only a 0.02% change of dry weight, hence the need for many applications.
10 HTI powder is High Temperature Insulation, a white refractory brick powder supplied by Macgregor and Moir Ltd, 27 Queensferry Street, Glasgow G5 0XJ; Tel: + 44 141 429 4294; Fax: +44 141 429 4288.
11 Formalin is the name given to formaldehyde in water. If it is added too quickly to a solution of casein and lime, a gelling action occurs, and a natural plastic film is produced. Minute solutions (e.g. 5 ml per 1 litre of wash) can inhibit the development of mould spots, but their effects are only transitory.
12 The friable surfaces of calcareous sandstones, such as Chilmark and Reigate stones, in England have apparently been strengthened by the use of lime watering. However, this practice is not recommended.
13 For further information on English Heritage's building materials research programme, and especially on projects AC1, AC2, AC5 and AC23, to which this current work AC9 relates, see Teutonico 1998 or write to Building Conservation and Research Team, English Heritage, 23 Savile Row, London WlS 2ET, UK; Fax: + 44 207 973 3249 or 3130; email: john.fidler@english.heritage.org.uk.

BIBLIOGRAPHY

Ashurst J and Ashurst N, 1988 *Practical Building Conservation*, English Heritage Technical Handbook Series **1**, Aldershot, Gower Technical Press.

Ashurst J, 1990 Surface treatments: Part 2 The cleaning and treatment of limestone by the lime method, in *The Conservation of Building and Decorative Stone*, (eds) Ashurst J and Dimes F, London, Butterworth Heinemann, 169–183.

British Standards Institution, 2000a *BS EN 459:2000 Building Limes*. London, British Standards Institution.

British Standards Institution, 2000b *BS 8221/2 British Standard Code of Practice The Cleaning and Surface Repair of Buildings*, London, British Standards Institution.

Brajer I and Kalsbeek N, 2000 Limewater absorption and calcite crystal formation on a limewater-impregnated secco wall painting, in *Studies in Conservation* **44**:3, 145–156.

Caroe M B, 1985 Wells Cathedral conservation of figure sculptures, 1975–1984, in *APT Bulletin* **17**:2, 3–13.

Caroe M B, 1987 Conservation of figure sculptures: Wells Cathedral 1975–1986, in *APT Bulletin* **19**:4, 11–15.

Church A H, 1862 *Improvements in the Means of Preserving Stone, Brick, Slate, Wood, Cement, Stucco, Plaster, Whitewash, and Colour Wash from the Injurious Action of Atmospheric and other Influences, etc*, 28th January, British Patent 220, London.

Church A H, 1904 *Copy of Memoranda by Professor Church FRS, furnished to the First Commissioner of His Majesty's Works, etc Concerning the Treatment of Decayed Stonework in the Chapter House, Westminster Abbey*, Parliamentary Command Paper 1889, HMSO, London.

Clarke B L and Ashurst J, 1972 *Stone Preservation Experiments*, Building Research Establishment Information Paper, Garston, Building Research Establishment.

Martin W, 1996 Stone consolidants – a review, in *A Future for the Past: Strategic Technical Research in the Cathedrals Grants Programme, Proceedings of a Joint Conference of English Heritage and the Cathedral Architects Association held in London 25th & 26th March 1994*, London, James & James (Science) Publishers Ltd, 30–50.

Powys A R, 1929 *The Repair of Ancient Buildings*, London, Dent.

Price C A, 1984 The consolidation of limestone using a lime poultice and limewater, in *Adhesives and Consolidants*, London, International Institute for Conservation (IIC), 160–162.

Price C A and Ross K D, 1984 The cleaning and treatment of limestone by the lime method Part II: A technical appraisal of stone conservation techniques employed at Wells Cathedral, in *Monumentum*, **27**:4, 301–312.

Price C A and Ross K D, 1990 Technical appraisal of stone conservation techniques at Wells Cathedral, in *The Conservation of Building and Decorative Stone* **2**, (eds) Ashurst J and Dimes F, London, Butterworth Heinemann, 176–184.

Price C A, Ross K D and White G, 1988 Further appraisal of the lime technique for limestone consolidation: Using a radioactive tracer, in *Studies in Conservation* **33**, 178–186.

Salzman L F, 1952 (1992) *Building in England down to 1540: A Documentary History*, Oxford, Clarendon Press.

Sampson J, 1998 *Wells Cathedral West Front: Construction, Sculpture and Conservation*, Stroud, Sutton Publishing.

Schaffer R J, 1932 (1972) *The Weathering of Natural Building Stones*, Building Research Special Report **18**, facsimile reprint, London, HMSO.

Teutonico J M (ed.), 1997 *The English Heritage Directory of Building Limes*, Shaftesbury, Donhead Publishing Ltd.

Teutonico J M (ed.), 1998 English Heritage's Research Programme: a schedule of projects on historic building materials decay and their treatment 1992/3–1997/8, in *Metals, English Heritage Research Transactions: Research and Case Studies in Architectural Conservation* **1**, James and James Science Publishers, London, 117–118.

Teutonico J M, McCaig I, Burns C and Ashurst J, 1994 The Smeaton Project: Factors affecting the properties of lime-based mortars, in *APT Bulletin* **25**: 2 & 3, 32–49.

Woolfitt C and Durnan N, 1995 (1996 revised edn) *Lime Method Evaluation: A Survey of Sites where Lime-Based Conservation Techniques have been Employed to Treat Decaying Limestone*, Unpublished report for English Heritage, Historic Building & Site Services, Department of Conservation Science, Bournemouth University, Bournemouth.

FURTHER READING

Ashurst J, 1983 The cleaning and treatment of limestone by the 'Lime method' Part I, in *Monumentum* **26**:3, 235–252.

Caroe M B and Caroe A D R, 1977 (1983) Wells Cathedral: The West Front Programme: interim report on aims and techniques, in *Transactions of the ASCHB*, **2**, 3–10.

Harris R, 1982 A survey of medieval repairs to the fabric of the West Front, Wells Cathedral, in *Transactions of the ASCHB*, **7**, 7–8.

Peterson S, 1981 Lime water consolidation in *Mortars, Cements and Grouts used in the Conservation of Historic Buildings: Proceedings of the ICCROM symposium, Rome, November 1980*, Rome, The International Centre for Studies in the Conservation and Restoration of Cultural Property (ICCROM), 53–61.

Price C A, 1981 *Brethane Stone Preservative* BRE Current Paper CP1/81, January, Garston, Building Research Establishment.

ACKNOWLEDGEMENTS

This paper has been developed from an oral presentation given at *Preservation Treatments for Historic Masonry: Consolidants, Coatings and Water Repellents*, an international symposium held in New York on 11th and 12th November 1994 which was organised by the Historic Preservation Education Foundation, the New York Chapter of the Association for Preservation Technology (APT) and the US National Park Service. Subsequently, the original paper was published (copyright English Heritage) in the *APT Bulletin*, **XXVI** (4), 1995, 50–56.

Photographs credited in the figure captions above to Jerry Sampson are held in joint copyright with the Chapter of Wells Cathedral whose kind permission to make reproductions here are gratefully acknowledged.

AUTHOR BIOGRAPHY

John Fidler RIBA is a chartered architect and Head of Building Conservation and Research at English Heritage, where he is responsible for developing technical policy and providing technical advice for research and development work on building materials decay and their treatment, for technical training and for the organisation's outreach campaigns and technical publications. He was the first Historic Buildings Architect for the City of London Corporation, the first national Conservation Officer for Buildings at Risk, and was the youngest and last Superintending Architect to maintain the country's historic estate. Fidler is the author of numerous technical publications on aspects of building conservation and was for ten years architectural editor of *Traditional Homes* magazine.

Lime method evaluation
A survey of sites where lime-based conservation techniques have been employed to treat decaying limestone

CATHERINE WOOLFITT

Ingram Consultancy, Netley House, Gomshall, Guildford GU5 9QA, UK
Tel: + 44 (0)1483 205170; Fax: + 44 (0)1483 205175

Abstract

This paper presents the results of a survey of the lime method commissioned by English Heritage in 1994 and completed with revisions in 1996. The term 'lime method' encompasses a range of lime-based conservation techniques, including cleaning, lime water consolidation, mortar repairs and shelter coat application. The English Heritage survey set out to determine the effectiveness of these various lime-based treatments applied to decaying limestone in buildings. Ten sites treated by the lime method from the 1970s to 1992 were surveyed. This paper presents observations on these sites, general conclusions regarding treatments, their effectiveness and durability and recommendations for future research.

Key words

Limestone, conservation, decay, survey, cleaning, repair, shelter coating, lime watering

PREAMBLE

BY NICHOLAS DURNAN

Lime-based conservation techniques for limestone are not new, but a considerable renewal of interest in their potential arose from the need to carry out extensive conservation work on some of England's major limestone cathedrals in the 1970s. Since then, lime-based techniques have developed considerably so that they now contribute significantly to limestone conservation in the United Kingdom and other European countries. A review of the past three decades is, by now, essential.

The site inspections and observations which formed the survey on which this paper is based have been supplemented by information provided by the conservators responsible for the work visited. Their comments indicate the great importance placed on the visual enhancement and intelligibility of sculpture and architectural detail, including the removal of any disfiguring accretions and interventions and the provision of visually harmonious repairs and protective treatments. These conservators invariably emphasise the need to develop hand and eye skills in the best masonry tradition and value objective science as an essential support to this development.

The lime method of conservation has evolved through prolonged trial and error experimentation in workshops and on site; this approach is the correct way to establish the right level of intimacy with the characteristics, condition and form of a wide range of stones and buildings.

There is little question that, in general, the lime method treatment significantly improves the appearance of a sculpture and, even more, the appearance of a whole façade. The visual gain is undoubtedly important and may be acceptable as an end in itself, but should not be confused with the effectiveness, or otherwise, of the method as a preservative measure. The long-term effectiveness of any treatment is far more difficult to assess than its appearance, but this survey goes some way towards such an assessment (Fig 1). [1]

Figure 1. *Wells Cathedral showing the appearance of sculpture on the West Front after shelter coating in the 1970s (Photograph by John Ashurst).*

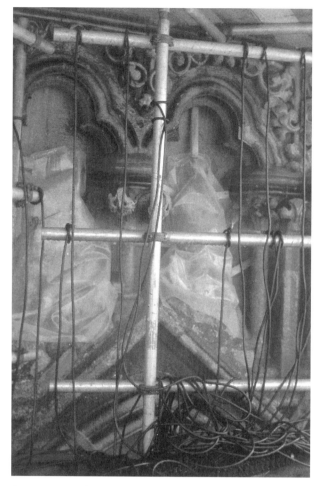

Figure 2. An early use of timed nebulous sprays at Wells Cathedral (Photograph by John Ashurst)

Figure 3. Albert Memorial. Cleaning of Portland limestone by Nimbus Conservation with timed nebulous sprays adapted from industrial water lines with articulated arms (Photograph by John Ashurst).

INTRODUCTION

By Catherine Woolfitt

In 1994 English Heritage commissioned a survey of the effectiveness of the lime method as a treatment for decaying limestone in buildings (Woolfitt & Durnan 1995). English Heritage and other public bodies commit substantial resources to remedial work programmes which include lime-based conservation methods. In the last thirty years, large-scale lime method programmes have been carried out at Wells, Exeter, Rochester and Canterbury cathedrals, among numerous other prominent sites. An assessment of the lime method was thought necessary to indicate which aspects of the method were successful and to assist specifiers and users, principally architects and conservators.

Conservators led by Professor Robert Baker and Eve Baker at Wells Cathedral developed and refined the lime method originally as a concerted approach to limestone conservation. Eve Baker was a wall paintings conservator and it seems that, under her influence, a number of wall-painting techniques, such as consolidation of lime plaster by lime watering, were adopted for limestone.

Experience in the use of lime for limestone conservation disseminated from Wells, and many of the conservators now practising lime-based techniques either trained at Wells or learned the method from a Wells conservator. The use of lime in a range of applications for limestones was a central part of what is sometimes termed 'the lime revival' (Holborow 1992). This resurgence in the use of lime focused on non-hydraulic lime putty, also known as fat lime or high calcium lime, to the exclusion of other forms of lime (Cowper 1927, Boynton 1980, BSI 2000). The lime revival evolved partly in opposition to cement-based repairs which have caused visible damage in recent decades. By contrast lime-based repairs are specifically intended to act sacrificially.

There are three main reasons for the widespread adoption of the lime method in England:

- There is an inherent trust in lime-based treatments, as the working properties of lime have been known for a long time and many lime treatment sites have been successfully retreated. Alternative surface treatments and fill materials, such as silane consolidants and resin composites, are less familiar, and their future behaviour, reversibility and potential for re-treatment are largely unknown, at least in the UK.
- Lime is widely perceived as being more compatible with limestone, than other repair materials, in terms of its physical properties and chemistry.
- There is a long tradition of the use of lime in the repair of buildings, and its present use in building conservation reflects a conscious attempt to maintain continuity of craftsmanship. Lime has been termed a 'low technology' material and conservators in this field tend to favour lime-based repairs for this reason. All the materials can be controlled, prepared, and even produced, by the conservator.

THE LIME METHOD

The lime method, briefly described by Fidler (this volume) encompasses three conservation applications of lime-based materials to limestone in buildings (Ashurst 1983, Ashurst 1990): after cleaning (Figs 2 and 3);

Table 1. Application of lime treatments to varying limestone surface conditions

Surface condition	Possible causes of deterioration/condition	Appropriate lime treatment and aim of treatment
Soot or gypsum crusts (black, unsightly crusts which obscure stone surfaces) (Fig 13)	Deposition of acidic atmospheric pollutants and other soiling materials; conversion of calcium carbonate to calcium sulphate	(Cleaning: to reveal hidden detail, to allow accurate assessment of condition and recording, to improve appearance, to remove potentially damaging soiling and salts and to allow stone to breathe), followed by: **mortar repairs** to protect vulnerable areas and allow obscured lines to read, **shelter coat** to protect from further weathering; also thought to inhibit further 'sulphation' and to act as a barrier to soiling
Exfoliating and blistering (gypsum skins or crusts) (Fig 5)	Conversion of calcium carbonate to calcium sulphate; wetting and drying cycles; salt crystallization; surface of skin may appear sound with decay apparent underneath	**Consolidation** with mortar slurries and repairs to strengthen and support weak surfaces; shelter coat – for reasons above
Friable and powdering surface (Fig 7)	Dissolution or loss of calcium carbonate binder; salt crystallization (various sources of water soluble salts); proximity to dense, impermeable repairs or pointing, often cementitious; often occurs beneath gypsum/soot crusts	**Lime water** for consolidation, where surfaces are valuable, historically; removal of decayed stone to sound substrate, where surface not original and not to be retained; possibly lime mortar repairs to protect surfaces and absorb salts, if present; shelter coat – for reasons above
Cracking and delaminating surface (Fig 7)	Conversion of calcium carbonate to calcium sulphate; salt crystallization; wetting and drying cycles; other stresses (possibly thermal)	**Mortar repairs** to fill cracks and voids, to strengthen stone, prevent moisture ingress and to absorb soluble salts, if present
Loss of stone	Any of the above or mechanical damage	**Mortar repairs** to suggest or restore lost detail, to protect vulnerable surfaces or areas; shelter coat – over mortar repair and repointing

consolidation, by lime watering; mortar repairs, using lime-based mortar fillings (this category includes repointing); and shelter coating, with lime-based shelter coats, blended by the conservator.

Much has been written about masonry cleaning processes associated with limestone conservation (Ashurst & Ashurst 1988, Ashurst 1994). A critical review of the lime treatment techniques and of attempts to scientifically evaluate them is set out in Fidler (this volume). Table 1 in this paper attempts to indicate how the condition of decayed stone determines which aspect of the lime method is usually applied. The headings in the table do not provide an exhaustive summary of limestone decay phenomena, but they and the associated treatments listed are indicative of common responses to these situations in the field.

Reports of effective consolidation by complete saturation of stones, for example by immersion in lime water, are fairly common, but this technique is not practical in most building contexts. The low solubility of lime in water is a problem as most of the pore space in the stone will be occupied by water; the amount of lime in solution available for deposition is minimal.

ENGLISH HERITAGE SURVEY METHOD

Survey sites

The survey was undertaken with a view to assessing the performance of lime-based treatments. One important consideration was the durability of lime treatments, the frequency with which, for example, shelter coats and mortar repairs need to be renewed. A principal objective was to identify maintenance cycles and requirements. Factors affecting the success and durability of treatments were to be identified and described.

Ten lime treatment sites were chosen for the survey in consultation with English Heritage. The specific objective was to survey ten sites treated in the past twenty years which were either English Heritage sites or had been grant-aided by English Heritage. Treatment records were available for all sites, either from English Heritage archives or from the conservators responsible for the work. The following sites were surveyed:

- Rochester Cathedral, the West Front and Cloisters (Durnan 1990, 1991a, 1991b and 1992)
- Canterbury Cathedral, the Romanesque Quire Arcade (Durnan nd)
- Wells Cathedral (Figs 4, 5 and 6) [2]
- Exeter Cathedral, Image Screen of the West Front (Durnan 1991c)
- Glastonbury Abbey, North and South Portals of the Lady Chapel (Alexander & Sampson 1986)
- Malmesbury Abbey, the South Porch [3]
- Chichester Market Cross (Odgers 1987, 16–17)
- Bath Abbey, the West Front (Odgers 1993)
- Basing Chapel of the Holy Trinity, Basingstoke (Marsh 1989)
- Old Gorhambury, St Albans, Porch of Sir Nicholas Bacon's house (Fig 7) (Ashurst 1979)

Site inspections were carried out by Catherine Woolfitt, with either Nicholas Durnan or Professor John Ashurst.[4] Where possible, the conservators responsible for the work were interviewed on site during the inspection. In addition to treatment reports, photographs of the sites prior to and during treatment were obtained and compared to photographs taken by Catherine Woolfitt or by Nicholas Durnan in the course of the survey.

A survey of literature on the lime method was conducted simultaneously with the site inspections. Searches were carried out on the on-line bibliographic database of the Conservation Information Network (CIN) and sources obtained are noted where relevant in this paper

Figure 4. Wells Cathedral, a heavily decayed sculpture prepared to receive mortar repairs (Photograph by John Ashurst).

Figure 5. Wells Cathedral, an exfoliating and blistering crust on sculpture and niche of Doulting limestone (Photograph by John Ashurst). See Colour Plate 9.

Figure 6. Wells Cathedral. Head of female figure with crack in nose. Projecting features on sculpture such as noses, toes and hands have larger surface areas subject to increased wetting and drying cycles, and are subsequently more prone to decay. Note also the cement capping on top of the head (Photograph by John Ashurst). See Colour Plate 10.

Figure 7. Old Gorhambury. Friable and powdering clunch (chalk) surfaces with limewash from the work programme of 1980 mostly lost (Photograph by Catherine Woolfitt). See Colour Plate 11.

(Zacharopoulou 1994). Although there is a tremendous amount of literature on the subject of lime generally in historic buildings, the volume of research and publications specifically addressing the lime method is relatively small.

Questions to be addressed: the survey form

At the outset it was necessary to establish some criteria for assessing the success of treatments. Any assessment must take into consideration the aim of treatments, as well as the initial condition of the stone. Criteria for an effective treatment may be grouped into three general categories.

Visual/aesthetic effect
A treatment may be based primarily on aesthetic considerations, with the intention of enhancing appearance rather than concern for the condition of the stone guiding decisions.

Conservation criteria
The aim of lime treatments is to reduce or arrest the rate of limestone decay and to protect stone from further weathering.

Repairing existing damage
This includes, for example, the removal of cement-based repairs and their replacement with lime-based mortars and pointing, and the remodelling of lost forms, where appropriate.

Broad questions which required answers and guided the design of the site survey form were:

- How durable are lime treatments, particularly shelter coats and mortar repairs?
- What maintenance cycles might be expected for shelter coats and mortar repairs?
- Do lime treatments significantly influence the rate of decay of treated stone?
- Does cleaning reduce the rate of limestone decay and increase the permeability of gypsum crusts, as some conservators believe?
- How can the above be measured and/or assessed?
- Given the large number of variables in a lime treatment site, including decay processes, stone type and condition, repair materials and techniques, is it possible to predict the behaviour of treated stone and the lime repair additions?
- Can lime-based treatments be improved by variations in formulation and application?
- Do certain types of limestone respond better to lime-based treatments than others?
- Is it possible and/or desirable to establish guidelines on the use of the lime method for architects and practitioners?

The survey form (Annex A), served as a guide rather than a rigid framework for making observations on site. It was difficult to anticipate the optimal layout of the form in advance and the lengthy survey form was found to be quite cumbersome. If this survey exercise were repeated, the form could be simplified and made more flexible. A separate form for recording data from treatment records would be useful, with a second abbreviated form for recording observations on site.

RESULTS OF THE SITE SURVEY

Table 2 summarizes information on the survey sites, including the limestones used in the buildings, the date of original stonework, the date of the most recent treatment and the architect and conservator responsible.

Purpose of treatments/criteria for success

Conservators and architects cited a number of considerations which guided treatment decisions. The following were factors in lime method programmes: the uncover-

Table 2. Survey sites

Site	Limestone type	Date of stonework	Treatment date	Architect/Conservator
Rochester Cathedral, cloisters and West Front	mainly Caen	twelfth century (and later replacement)	Cloisters 1989–92, West Front 1991	Martin Caroe/Nick Durnan
Canterbury Cathedral, Quire Arcade	Caen and Quarr	twelfth century	1986–8	Peter Marsh/Nick Durnan
Wells Cathedral, West Front	mainly Doulting	thirteenth century (and later replacement)	Various 1970s–1990s (including retreatments)	initially Caroe and Martin Architects/Professor Robert Baker
Exeter Cathedral, West Front	Beer (with later Ketton and Bath)	fourteenth-fifteenthth century	1992 (retreatment of 1978–85 work)	Peter Bird/Nick Durnan
Glastonbury Abbey, Lady Chapel Portals	figure sculpture of Dundry; jambs of Doulting, Bath, Chilcott	twelfth century	1986	Alan Rowe/Hebe Alexander
Malmesbury Abbey, South porch	Bath, quarried from Box Hill	twelfth century	1983 (retreatment of 1977 work)	John Ashurst/Richard Marsh
Chichester Market Cross (NE, N, NW elevations)	Caen	1501	1986	A Bridges, Building Surveyor/David Odgers and Geoffrey Preston
Bath Abbey, West Front	Bath - mainly Box	early sixteenth century with much later	1991: top half, 1992: bottom half	Martin Caroe/David Odgers
Basing Chapel of the Holy Trinity, Basingstoke	Chalk (clunch)	early sixteenth century	1989–92	Richard Marsh Conservator
Old Gorhambury, St Albans	mainly Totternhoe (clunch)	1563–1570s	1980	John Ashurst/Bob Bennett

Figure 8. Henry VII's Chapel, Westminster Abbey in 1976 before cleaning and surface treatments (Photograph by John Ashurst).

Figure 9. Henry VII's Chapel in 1996 after cleaning and surface repair and treatment, including shelter coating. The appearance of the stonework in Figure 8 has been considerably altered (Photograph by John Ashurst).

ing and recording of detail, improvement in appearance, removal of previous damaging repairs (almost always cement-based fillings or repointing) and the conservation of valuable carved surfaces.

Recording of freshly cleaned sculpture was part of the conservation work at Rochester, Canterbury and Wells cathedrals, and at Glastonbury Abbey, Bath Abbey and Old Gorhambury. Removal of thick black crusts from the sculpture permitted recording by drawing and photography. The architects and archaeologists involved at these sites believed that cleaning and recording were essential before further detail was lost to limestone decay. In addition, the removal of black soiling crusts was thought to be beneficial for limestone surfaces, eliminating a source of potentially harmful salts and increasing the permeability of the surface beneath.

The removal of older, damaging repairs and replacement in lime-based repairs typically features in conservation programmes on sites which were maintained in the past by patch fillings and repointing in cement-based mortar. This was one explicit motive for the recent lime method programmes at Rochester, Wells, Chichester and Bath. These aims, to clean and remove potentially harmful soiling and old repairs and to record newly exposed detail, are quite readily identified and achieved.

It is more difficult to define some other aims, such as aesthetic effects, and determine whether or not they were achieved. A visual improvement is part of the objective of the treatments at all sites, but was given more weight at some sites than others. For example, a shelter coat was sometimes used to harmonise large areas of stonework and to integrate new repairs and replacement stone with existing stonework. This was the case at Rochester, Exeter, Chichester, Bath Abbey, and, since this survey was carried out, at Winchester Cathedral and Henry VII's

Figure 10. Wells Cathedral. An example of mortar repairs used to restore form; the left side of the helmet has been renewed in lime mortar (Photograph by John Ashurst).

Figure 11. Romanesque Quire Arcade, Canterbury Cathedral. Shelter coat was used sparingly in the conservation of sculpture in the Arcade. The front of this capital was shelter coated, but the fine carving revealed by cleaning of the return faces of the capital was left untreated (Photograph by Catherine Woolfitt). See Colour Plate 12.

Chapel, Westminster Abbey, sites outside the scope of the present survey. In these cases shelter coats were intended to play a dual role, that of a sacrificial, protective layer and a surface coating for visual effect (Figs 8 and 9).

Mortar repairs are also typically intended to fulfill the dual function of enhancing appearance and behaving sacrificially to the stone. Mortar repairs were often employed to help read lost forms and to suggest, or even restore, lost shapes or lines. This aesthetic function was sometimes of primary importance in mortar repairs to sculpture and architectural detail at Wells, Exeter, Chichester and Bath (Fig 10). On the other hand, the enhancement of overall appearance was subordinate to conservation concerns at Canterbury and Glastonbury. In contrast to the other sites, shelter coats and mortar repairs were used selectively on the Romanesque Quire Arcade at Canterbury and the Lady Chapel Portals at Glastonbury (Figs 11 and 12). The early date and fine quality of sculpture at these two sites required a particularly conservative approach.

It should be noted from Table 2 that Wells, Exeter and Malmesbury are re-treatment sites. This survey clearly indicated that maintenance of lime-based repairs is essential, especially of mortar fillings which are often designed to fail in order to protect the stone in which they are placed. Malmesbury Abbey has a specific maintenance problem, decay by soluble salts, caused by a residue of

Figure 12. North Portal of the Lady Chapel, Glastonbury Abbey. The use of water was minimal in the conservation of the North Portal to prevent saturation and softening of the stone and mobilization of salts present in the joints. This consideration and the desire not to obscure the sculpture surfaces in any way led to a decision not to shelter coat the arch (Photograph by Nicholas Durnan). See Colour Plate 13.

potassium salts, deposited when the south porch was used for storing gunpowder. Shelter coating and dental repairs to five trial stones on the south porch were an attempt to draw out the salts, leaving the limestone surface intact (Figs 13 and 14).

Figure 13. The South Porch of Malmesbury Abbey in 1995. Five stones were selected for trial shelter coating in 1983 (Photograph by Catherine Woolfitt).

Figure 14. Detail of shelter coated stone in Malmesbury Abbey South Porch. Salts showing as white patches have reappeared through the shelter coat and mortar repairs in the same areas where they were recorded in the pre-treatment survey (Photograph by Catherine Woolfitt). See Colour Plate 14.

Results of survey: site observations/conditions method

Table 3 presents a summary of the application of lime method techniques and the present condition of stone surfaces at each survey site. The original report prepared for English Heritage included detailed observations and treatment information for all sites. This has been condensed into the table and text.

Cleaning

The choice of cleaning method at the various sites was determined by the condition of the limestone surface, and the depth and type of soiling. The presence of water-soluble salts at Glastonbury and Malmesbury abbeys, visibly disrupting stone and joint surfaces, was, for example, an additional factor to be considered.

The cleaning methods used at the survey sites were limited, reflecting the small number of methods currently recommended for limestone. Water sprays and brushing, using hand-held sprays or timed mist sprays, micro-air abrasive units, ammonium carbonate poultices [5] and hot lime poultices were used, often in combination on a single facade to accommodate varying surface conditions. Air-abrasive pencil was used on dense crust soiling on delicate and valuable carving where controlled and careful cleaning was required. Poultices with paper pulp, clay, carboxymethyl cellulose and ammonium carbonate were often inadequate for heavily encrusted surfaces. Air abrasive cleaning, typically with fine aluminium oxide (less than 150 microns), was often the safest and most controllable method for removing dense crusts, although it is also the most time-consuming technique. It is certainly the best method for cleaning stone when there is potential for the survival of polychromy, which is often the case with medieval sculpture. The use of hot lime poultices on sculpture, once an integral part of the lime method, has been largely supplanted by other cleaning methods. The lack of evidence for the consolidating effect of the hot lime poultice, the lengthy dwell times, the potential for damage to delicate surfaces from the weight and reactivity of the putty, and inherent

Figure 15. North Portal of the Lady Chapel, Glastonbury Abbey. The sculpture of the arch was dry cleaned with air abrasive in 1986. Some resoiling is now occurring (Photograph by Nicholas Durnan).

Table 3. Summary of assessment of lime method techniques at survey sites

Site	cleaning method	lime watering	mortar repairs: general observations on use/current condition	shelter coat	present condition of shelter coat
Rochester Cathedral, Cloisters and West Front	ammonium carbonate poultice (sculpture and detail), water washing	40 applications to all Caen stone	all exfoliating edges, cracks and blisters on original surfaces filled; overall mortar repairs sound, but 'feather edges' fail	over all West Front and Cloisters	much lost on flat, rain-washed surfaces. Survives in sheltered areas and recesses, in niches, tool marks and joints. Effect of facade unified by shelter coat beginning to fade
Canterbury Cathedral, Quire Arcade	ammonium carbonate poultice (minimal) and air abrasive pencil for sculpture/detail. Timed mist sprays for ashlar and arcade	40 applications to all surfaces, except plain ashlar	consolidation prior to cleaning in some cases with lime mortar, otherwise relatively minimal use	minimal use on faces of capitals and shafts	about 50% survives on sheltered capitals. Mostly weathered off (gone on) exposed shafts
Wells Cathedral, West Front Image Screen	various, including hot lime poultice and air abrasive pencil	various	vary greatly in successful matching of colour and texture and in modelling, due to various conservators	over all West Front	survives only in sheltered areas. Orange colour from Hornton Blue stone dust
Exeter Cathedral, West Front	portable mist sprays	40 applications to all stonework	decayed mortar repairs from previous repair programme replaced; New fillings to suggest lost forms; Approx 10–15% of repairs decaying, mainly on S. end where previously capped in cement	over all West Front	partial survival overall, greater loss on weathered surfaces; survival on Ketton more noticeable. Overall effect of shelter coat unifying facade persists
Glastonbury Abbey, Lady Chapel portals	various, including air abrasive (north portal) and water (south portal)	north portal – below capitals only, south portal – 20 applications	pre-consolidation (prior to cleaning) with mortar slurries; Mortar repairs to support areas, fill cracks and eliminate water traps; More rapid weathering of stone and mortar on south portal	on south portal hood mould only (more exposed to weathering than north portal)	mostly weathered from hood mould
Malmesbury Abbey, south porch	minimal cotton wool water poultice desalination	not used	minimal fillings to support flakes and fill cracks; Salts have re-appeared where they were noted before treatment	5 trial stones only on door of south porch	some survival on stones sheltered by arch. Orange staining from Hornton Blue stone dust
Chichester Market Cross (NE, N, NW elevations)	timed mist sprays	40 applications to all stonework	mortar repairs to rebuild mouldings and integrate new carving with existing stonework	all elevations	patchy survival, especially in sheltered areas. Does not survive on earlier cement repairs and no longer masks difference between original Caen and Lepine replacement
Bath Abbey, West Front	various, including ammonium carbonate poultice and water	for friable stone	quality of repairs varies, depending on conservator; Some sculptures re-modelled in mortar from evidence of nineteenth-century plaster casts	over all West Front	partial survival overall; much weathered at ground level. Overall effect of facade unified by shelter coat beginning to fade
Basing Chapel of the Holy Trinity, Basingstoke	removal of surface flakes, brushing	all stonework	repairs to cap eroding surfaces, adhere lifting flakes. Many repairs lost with flaking stone	over all surfaces	lost with surface flakes. Where surface survives, some shelter coat still adheres (over approx. 50% of area)
Old Gorhambury, St Albans	various, including hot lime poultices, air abrasive pencil, murasol biocide and brushing off surface flakes	40 applications	mortar repairs to clunch and various harder decorative stones; Repairs to decorative stones survive. Problem with ongoing decay of clunch	2 coats of lime wash over all clunch surfaces, except trial consolidation frieze	limewash lost with flaking clunch surface. Most of limewash and stone surface lost on west sides; loss slightly less severe on east side

Figure 16. Romanesque Quire Arcade, Canterbury Cathedral. Conservation work in 1986 to 1988 included the installation of a protective projecting roof, cleaning and selective mortar repairs and shelter coating (Photograph by Catherine Woolfitt).

lack of control in this method have made its use almost obsolete on sculpture (for which it was originally designed).

In general, the approach to cleaning has become more conservative. The amount of water used and wetting times are minimized to avoid softening surfaces and mobilizing salts. For example, the use of water was kept to a minimum in the cleaning and consolidation of the sculpture on the North Portal of the Lady Chapel at Glastonbury (Fig 15). Dry cleaning was carried out with an air abrasive unit in order to prevent the movement of salts from the joints into the stones. Shelter coating, which requires pre-wetting of surfaces, was carried out on the hood mould of the South Portal only. It was not always considered essential to remove all soiling, particularly if it appeared to be stable. At Canterbury traces of dark soiling on the flat arcade which were not causing decay were left as their removal could potentially have been damaging, removing original tool marks (Fig 16).

Cleaning at Glastonbury Abbey and Exeter Cathedral permitted detailed examination of sculpture, yielding important information for conservators, art historians and archaeologists concerning the style, date and polychromy, which was subsequently published (Sampson 1992, Sinclair 1992).

Consolidation by lime watering
Lime watering was an integral part of the conservation programmes at all sites surveyed, with the exception of Malmesbury Abbey. With the single exception of Glastonbury Abbey North Portal, lime watering was always followed by shelter coating and, in the case of Old Gorhambury, by lime washing. This makes it virtually impossible to assess the effect of lime watering in isolation on limestone surfaces at the survey sites. Most conservators stated that lime watering caused some perceptible tightening of the surface.

While this treatment may be effective on lime plaster, and even limestone in an indoor environment, it seems from field assessments that the effect of lime watering is negligible on external stonework, subject to regular rain washing and weathering. Alternative consolidants to lime water were not used, or even considered, at any of the survey sites. Lime watering appears to have persisted in

Figure 17. Mortar repairs to this moulding and capital are poorly modelled, deviating from the lines of the original (Photograph by Catherine Woolfitt). See Colour Plate 15.

Figure 18. The mortar in this repair, situated at ground level on a prominent West Front, does not match the stone and contains large, visually obtrusive aggregate. A small shrinkage crack is also visible at the stone-mortar interface (Photograph by Catherine Woolfitt).

use on external stonework due to tradition rather than convincing evidence of its effectiveness.

Mortar repairs
In spite of the many variables at the ten sites, such as stone and decay types, treatment date, repair materials and methods, and conservators, a few general observations can be made. First, the skill and experience of the conservator are of paramount importance in achieving a mortar repair which is successful, both aesthetically and

in terms of durability. Skill in colour and texture matching, placing, modelling and conditioning/curing of the mortar filling, all contribute to a visually acceptable and sound repair. Artistry and sensitivity to the style of the subject are particularly important in repairs to sculpture and in instances where the repair is also a reinstatement or suggestion of lost detail. In these cases the repair should maintain the fluency of the original contours. The author occasionally observed that, where this was not the case, mortar repairs were out of plane, relative to the original stone, and poorly modelled (Figs 17 and 18).

The condition of stone surfaces prior to treatment was, as anticipated, a critical factor in the condition of mortar repairs. Mortar repairs to heavily decayed stone surfaces, for example those affected by salts associated with old cement repairs, were much more prone to failure than repairs to relatively sound stone. This was observed at Exeter and Wells Cathedrals and Malmesbury Abbey, in particular. If the stone was initially very decayed, the author observed generally that it might continue to decay around the mortar repair and may eventually rupture, pushing the repair off the surface. Mortar repairs were not, however, observed promoting decay of comparatively sound adjacent stone at any survey sites.

Conservators consistently stated that mortar repairs should be porous and moisture-permeable to draw soluble salts and moisture from limestone. At the same time some degree of durability is desired to minimize maintenance of repairs. In general, open and coarse-textured mortars are assumed to be more durable than fine-textured mortars. Decayed limestone is thought to absorb water more readily than sound limestone. Ideally the mortar repair should absorb salts and moisture from the weakened stone and weather sacrificially in its place. Some of these observations are soundly based on practical experience, but there are assumptions about the physical properties of lime mortars, relative to limestones, which have not been adequately verified and tested

Mortar repair mixes were not tested, relative to the original stone, at any of the survey sites. Sample mortars were prepared and chosen on the basis of appearance, texture and colour. The effects of aggregate type (eg porous versus non-porous) and grading, and binder to aggregate ratio on physical properties, such as porosity and moisture permeability also need to be established. The Smeaton project has identified factors which influence the strength and durability of mortars (Teutonico et al 1994). These factors need to be investigated and defined in the context of mortar repairs to decayed limestones where the performance criteria are different from those applicable to building mortars in general.

Shelter coating

Shelter coating was used both for visual effect and as a protective coating at the sites surveyed. At five of the sites (Rochester, Wells, Exeter, Chichester and Bath) a shelter coat was intended to unify stonework on a facade, to mask the differences between old replacement or original stone and new replacement or mortar repairs. At Bath Abbey and Rochester Cathedral sculptures on the West Fronts were accentuated by varying the pigmentation in the shelter coat,

Figure 19. Rochester Cathedral, West Front. Figure sculptures on the West Front were treated with different shelter coats and still appear a slightly different colour from the rest of the facade (Photograph by Catherine Woolfitt).

with the intention of enhancing the visual effect of the treated sculpture (Fig 19). The intended function of shelter coats at Canterbury, Malmesbury and the Basing Chapel was more purely protective: at these sites shelter coats were primarily sacrificial coatings.

The durability of shelter coats was, in general, surprisingly short. Shelter coats inspected at the survey sites had all weathered considerably, with overall erosion from exposed areas within a period of only three to four years. Shelter coats survived in joints, surface recesses and in sheltered areas, such as under projecting string courses and niches, for considerably longer periods, sometimes in excess of eight to ten years (Fig 20).

Weathered shelter coats are typically powdery to the touch and sometimes weather irregularly with whiter patches occurring, presumably where more lime survives than stone dusts and other pigmentation (Fig 21).

The ability of shelter coats to prevent the conversion of limestone to calcium sulphate and to act sacrificially by absorbing soluble salts has not been tested or proven.

CONCLUSIONS

Cleaning

Cleaning has been a visual success at the sites surveyed in terms of an improved appearance of sculpture and fa-

Figure 20. Rochester Cathedral, West Front. Detail of weathered shelter coat, showing survival of lighter material in joints and on north-facing side, with almost complete loss on southern exposure (Photograph by Catherine Woolfitt).

Figure 21. Chichester Market Cross with weathering surfaces supporting green organic growth

cades. In addition, cleaning of sculpture and detail has revealed previously obscured forms and permitted detailed recording and thorough assessments of surface conditions. The use of limited amounts of water seems to be desirable for deteriorated limestone but some gypsum crust surfaces are susceptible to softening with the application of water. It seems that in some cases the removal of soot crusts may be justified as a conservation measure to protect the stone beneath. Where the condition of the stone under the gypsum crust is friable, powdery or otherwise actively decaying, cleaning is beneficial. It is, however, more difficult to justify cleaning on conservation grounds when the gypsum crust adheres to sound stone, although cleaning will certainly be beneficial aesthetically. There is concern that the reduction of gypsum skins by any cleaning method may result in loss of original, albeit altered, surfaces.

Although the mechanism of the conversion of calcium carbonate to calcium sulphate is generally understood, the various ways in which different limestones behave has not been entirely explained. The chalk at Old Gorhambury and the Basing Chapel had deteriorated more severely and rapidly than the other stones surveyed and in a different way. The chalk typically failed by detaching in many small flakes, a mode of failure relating presumably to the microporous structure of the limestone (see Fig 11). It is possible to analyse limestone surfaces for calcium sulphate content, but the difficulty lies in knowing how much calcium sulphate conversion various limestones can tolerate before they begin to decay.

Special care must be taken in the selection of method and cleaning of carved medieval surfaces which may bear traces of polychromy. A pre-treatment survey of limestone surface conditions, which may vary considerably across a single building facade, should precede any cleaning and repair programme.

Laser cleaning has been omitted from the discussion at this point, since lasers were not used at any of the sites surveyed. The use of lasers to clean stone in buildings is a relatively recent development in the UK, and to date has mostly been confined to sculpture (Beadman & Scarrow 1998, Cooper 1998). This situation could change in the future with developments in laser cleaning technology and wider familiarity with laser cleaning equipment. At present laser cleaning is comparable to the air-abrasive pencil in terms of the rate of cleaning, and its use is accordingly impractical on larger surface areas of buildings. The laser is a safe and effective tool in experienced hands, but, like other cleaning methods, it can cause damage if not used correctly without proper experience and training. Polychromed surfaces must be treated with particular care since laser radiation can alter some pigments (Shekede 1998).

Lime watering

Although lime watering continues to be specified and used for the consolidation of friable limestone surfaces,

Plates

Plate 1. **Brethane:** *Detail of the base of the loggia pier at Berry Pomeroy Castle (1997) (English Heritage Photo Library). See Figure 4, page 7.*

Plate 2. **Brethane:** *Treated block of Hardwick sandstone at Bolsover Castle, SW entrance, treated in 1982 (English Heritage Photo Library). See Figure 5, page 8.*

Plate 4. **Brethane:** *Efflorescence on the surface of treated areas at Kenilworth (1997) (English Heritage Photo Library). See Figure 12, page 11.*

Plate 3. **Brethane:** *Untreated block of Hardwick sandstone at Bolsover Castle, SW entrance (English Heritage Photo Library). See Figure 6, page 8.*

Plates

Plate 5. **Lime Treatments:** *Limestone decay. Sulphates and ruptured gypsum crusts are in the foreground with cavernous powdering decay behind (Photograph by Jerry Sampson). See Figure 8, page 23.*

Plate 6. **Lime Treatments:** *The lime technique. Holes are drilled for access for flushing out harmful salts and for the introduction of grout and mortar dentistry repairs, as a precursor to lime watering and shelter coating. Only a comprehensive approach to the lime technique, with ongoing maintenance, will succeed. Sometimes the treatment is not as gentle as is claimed (Photograph by Jerry Sampson). See Figure 9, page 23.*

Plate 7. **Lime Treatments:** *Lime watering. Many applications are necessary before the calcium hydroxide in solution appears to have any effect, though scientific testing has still to prove that it actually works. It is possible to get the same surface hardening by using tap water (Photograph by Bill Martin, English Heritage). See Figure 10, page 24.*

Plate 8. **Lime Treatments:** *Lime shelter coat. The long-term maintenance of the shelter coat is a prerequisite of good practice (Photograph by Bill Martin, English Heritage). See Figure 11, page 25.*

Plate 9. **Lime evaluation:** *Wells Cathedral, an exfoliating and blistering crust on sculpture and niche of Doulting limestone (Photograph by John Ashurst). See Figure 5, page 32.*

Plate 10. **Lime evaluation:** *Wells Cathedral. Head of female figure with crack in nose. Projecting features on sculpture such as noses, toes and hands have larger surface areas subject to increased wetting and drying cycles, and are subsequently more prone to decay. Note also the cement capping on top of the head (Photograph by John Ashurst) See Figure 6, page 32.*

Plate 11. **Lime evaluation:** *Old Gorhambury. Friable and powdering clunch (chalk) surfaces with limewash from the work programme of 1980 mostly lost (Photograph by Catherine Woolfitt). See Figure 7, page 32.*

Plates

Plate 12. **Lime evaluation:** *Romanesque Quire Arcade, Canterbury Cathedral. Shelter coat was used sparingly in the conservation of sculpture in the Arcade. The front of this capital was shelter coated, but the fine carving revealed by cleaning of the return faces of the capital was left untreated (Photograph by Catherine Woolfitt). See Figure 11, page 35.*

Plate 14. **Lime evaluation:** *Detail of shelter coated stone in Malmesbury Abbey South Porch. Salts showing as white patches have reappeared through the shelter coat and mortar repairs in the same areas where they were recorded in the pretreatment survey (Photograph by Catherine Woolfitt). See Figure 14, page 36.*

Plate 13. **Lime evaluation:** *North Portal of the Lady Chapel, Glastonbury Abbey. The use of water was minimal in the conservation of the North Portal to prevent saturation and softening of the stone and mobilization of salts present in the joints. This consideration and the desire not to obscure the sculpture surfaces in any way led to a decision not to shelter coat the arch (Photograph by Nicholas Durnan). See Figure 12, page 35.*

Plate 15. **Lime evaluation:** *Mortar repairs to this moulding and capital are poorly modelled, deviating from the lines of the original (Photograph by Catherine Woolfitt). Figure 17, page 38.*

Plates

Plate 16. **Graffiti:** *Photograph of SGB1 applied to Monks Park limestone, clearly showing the depth of penetration and the structure of the barrier over the surface (BRE for English Heritage). See Figure 10, page 49.*

Plate 17. **Graffiti:** *SGB3 stained black with iodine on a limestone surface viewed in cross-section. Field of view 2x3 mm (BRE for English Heritage). See Figure 11, page 49.*

Plates

Plate 18. **Graffiti:** *Samples of Portland limestone treated with SGB2 (left), SGB1 (middle) and untreated (right) being subjected to the crystallization test (BRE for English Heritage). See Figure 17, page 52.*

Plate 19. **Graffiti:** *Efflorescence observed on samples of Portland limestone treated with SGB2 (left), SGB1 (middle) and untreated (right) after three days of the crystallization test (BRE for English Heritage). See Figure 18, page 52.*

Plate 20. **Graffiti:** *The final condition of samples of Portland limestone treated with SGB2 (left), SGB1 (middle) and untreated (right) after two weeks of the crystallization test (BRE for English Heritage). See Figure 19, page 52.*

Plate 21. **Graffiti:** *The result of high-pressure hot water cleaning from sacrificial graffiti barrier treated limestone samples (SGB1, top, and SGB3, upper middle) and weathered limestone samples (SGB1, lower middle, and SGB3, bottom). Green spray paint and black felt marker pen were used as graffiti types. The control samples were not treated with a barrier and were cleaned with high pressure water only (BRE for English Heritage). See Figure 20, page 53.*

Plate 22. **Graffiti:** *Abrasive damage to stone as a result of increased pressure. 20 bar (300 psi), 40 bar (600 psi), 60 bar (900 psi), 80 bar (1200 psi) (BRE for English Heritage). See Figure 21, page 53.*

Plate 23. **Soft wall capping:** *Hard wall capping at Hailes Abbey (Photograph by Chris Wood, English Heritage). See Figure 1, page 60.*

Plate 24. **Soft wall capping:** *Soft wall capping at Hailes Abbey (Photograph by Heather Viles). See Figure 2, page 60.*

Plate 25. **Bristol Temple Church:** *Fractured reveals showing the distinctive 'fire-reddening' which is caused by excessive heat on the iron-rich minerals within the stone. See Figure 4, page 79.*

Plate 26. **Bristol Temple Church:** *Fracturing along the line of the bedding planes caused by rapid cooling of stone following the fire. Frost has exacerbated these cracks over the years. See Figure 12, page 82.*

Plate 27. **Bristol Temple Church:** *The large ceramic dowels are evident just behind the nosing to the tracery which has now gone. See Figure 15, page 83.*

Plates

Plate 28. **Bristol Temple Church:** *A more dramatic example of the failure of this early repair where the dowels now stand proud and the nosing has become completely detached. See Figure 16, page 83.*

Plate 30. **Bristol Temple Church:** *Clay cups with oversized twisted spiral pins in each hole. See Figure 28, page 89.*

Plate 29. **Bristol Temple Church:** *Holes had to be drilled to allow thin copper wire to be threaded through and around the stone to prevent detachment during the repair. See Figure 19, page 85.*

Plate 31. **Bristol Temple Church:** *The latex rubber protection being removed after the resin had cured. See Figure 31, page 90.*

Plates

Plate 32. **Bristol Temple Church:** *The 1990 work was 'overfinished' producing a rounded appearance rather than the dramatic, sharp, angular appearance more typical of fire-damaged fractures. See Figure 32, page 91.*

Plate 34. **Bristol Temple Church:** *Nimbus Conservation Ltd used silicone around the holes to prevent resin spillage, rather than latex rubber. See Figure 36, page 92.*

Plate 33. **Bristol Temple Church:** *The 1991 work was far more successful at reproducing the original appearance of the damaged masonry. See Figure 33, page 91.*

Plate 35. **Bristol Temple Church:** *Detail of the 1991 repair taken eight years later, still looking satisfactory although as expected there has been some loss of shelter coat. See Figure 38, page 94.*

Plates

Plate 36. **Wellington Arch:** *View of the crack on the north side of the column (Photograph by Les Ayling, English Heritage). See Figure 2, page 98.*

Plate 38. **Wellington Arch:** *The completed repair (Photograph by Les Ayling, English Heritage). See Figure 10, page 104.*

Plate 37. **Wellington Arch:** *Close-up of the stone insert (Photograph by Les Ayling, English Heritage). See Figure 8, page 103.*

and the 'tightening of lime watered surfaces' is frequently cited, there is little evidence for its beneficial effect on external limestone surfaces. Shelter coating follows lime watering in the traditional lime method of application so that assessment of lime watering in isolation is virtually impossible. The effect of lime watering on weathered and rain-washed surfaces seems to be negligible and its use questionable, although there may be potential for its use on internal stonework, subject to further testing to confirm beneficial effects.

Mortar repairs

In general, mortar repairs were the most beneficial and durable of the lime method techniques in preventing loss of limestone surfaces and detail. The success of the mortar filling, in protecting delicate stone surfaces or in restoring or suggesting lost forms and lines, is dependent to a large extent upon the skill of the conservator in blending a good mortar mix and modelling an aesthetically sensitive repair. Even well executed repairs are likely to fail on very deteriorated stone, such as that contaminated with salt associated with previous cement repairs. Desalination prior to placing mortars is an option to be considered. The relationship between the physical characteristics of the stone and the mortar repair is understood through the intuition and experience of the conservator rather than analysis of properties such as porosity and permeability.

The variables in repair mixes, including the effect of aggregate types on physical properties, the effect of binder : aggregate ratios and the properties of other lime binders, such as hydraulic limes, have not been explored or tested. In all cases cutting out cementitious mortar pointing and fillings, and replacement with lime-based mortar, was beneficial aesthetically and appeared to reduce damage at the repair interface with the stone.

Shelter coating

Lime shelter coats weather severely within two to three years on exposed surfaces, with increased durability on sheltered areas. Sculptures treated with shelter coats are often situated in niches or arches so that the shelter coat at least provides more lasting protection in areas where it is most needed. Conservators generally believe that shelter coats afford limestone protection from weathering and ongoing conversion to calcium sulphate, although these assumptions have not been tested or proven. In general, weathering of shelter coats was more severe on south-facing elevations, for example at Rochester, Wells and Exeter.

Shelter coats were often used for aesthetic reasons, to unify areas of stonework or whole facades. Shelter coats were sometimes intended to facilitate later cleaning by acting as a barrier between soiling and the original stone surface and to prevent the formation of soot crusts. In theory shelter-coated surfaces should be easier to clean than untreated stone.

It was not possible to establish whether or not certain stones were more suitable for shelter coating and lime method treatment in general. An exception is that a shelter coat appears to bind poorly to very fine-grained limestones, such as chalk or Blue Lias; this was noted at the Basing Chapel and Old Gorhambury. Performance of shelter coats may be partly related to the texture of the stone, with a better bond forming on more coarsely textured limestones. In some cases shelter coats must be applied selectively. For example, they should not be applied to residues of original surface finishes, such as surviving polychromy. Shelter coats do not adhere properly to cement-based repairs.

General conclusions

In the course of this survey the assessment team asked conservators why the lime method was used and was told that other options were not considered. The lime method was used because it is traditional and because of its perceived success at Wells.

The prevailing attitude regarding the lime method (particularly lime watering and lime shelter coating) is to simultaneously acknowledge that these lime techniques may not have a measurable effect and to point out that at least they are reversible in the sense that they weather away. Lime-based interventions will not have unforeseen detrimental effects on the treated stone in the future. Synthetic or organic consolidants, such as silanes or resins, are viewed with distrust as irreversible treatments. The use of lime-based techniques in combination with other treatments has not been explored. For example, pre-consolidation with a silane followed by shelter coating may be an option in certain circumstances.

It is difficult to confirm on the basis of this survey that the rate of limestone decay will be reduced by a lime treatment, although mortar repairs will help retain limestone surfaces in a precarious condition and prevent further surface loss. An improved appearance certainly resulted, particularly of sculpture once obscured by heavy black soiling. Removal of soiling crusts is thought to enhance the condition of limestone surfaces, and certainly enhances appearance, but it is difficult to quantify physical effects.

It is also difficult to predict the future behaviour of treated stones, due to the many variables involved, but it may be possible to predict to some extent the behaviour of mortars whose constituents can be controlled. Research is required into the effects of various aggregates, such as sands versus crushed limestone (porous aggregate), the use of hydraulic lime binders and other variables in mortar repair formulations. Testing should focus on physical properties, such as porosity, permeability and the stone-mortar repair interface.

Lime method techniques, and particularly mortar repairs, have been successful in retaining architectural detail, but stonework must be inspected at regular intervals so that remedial works can be carried out. Although the lime method is safe, it should be borne in mind that treatments cost money, sometimes public money. If treatments are questionable and appear not to perform their intended function, which is the case generally with lime watering and occasionally with shelter coating, it is necessary to identify the fact and divert resources to other activities.

RECOMMENDATIONS FOR FUTURE RESEARCH

Lime watering

Further investigation into lime water application techniques may be of interest. It is difficult to achieve adequate depths of consolidant penetration with traditional spray application. It may be possible to improve penetration either by adding substances to increase the solubility of the lime [6] or by saturation application to avoid the usual problem of intermittent evaporation of water carrying the consolidant back to the outer stone surface.

Mortar repairs

Well-designed and properly placed mortar in the form of dental repairs plays a major and established part in the conservation of limestone sculpture and architectural detail. In this context 'well designed' means a mortar which is compatible with the stone in physical properties such as porosity, permeability, grain size and colour, but which will tend to behave sacrificially with exposure to prolonged wetting and drying.

One purpose of mortar research in relation to the lime method is the development of mortar mixes which have the same, or very similar, physical properties to the stone in which they are placed, or which will behave in a desired way, for example, sacrificially to the stone.

Field experience and research have already provided a substantial amount of information relating to mortar design, but the behaviour of mortar at the crucial stone-mortar interface has not been explored. This stone-mortar interface is generally the site of problems, such as cracking, trapping of moisture and discolouration. It is necessary to develop material and technique specifications to encourage moisture and salt migration into the mortar. Techniques of preparation and placement, variations in binder and aggregate specification, and methods of compaction and curing need to be compared in terms of controlling the migration patterns.

To control the number of variables, it may be advisable to begin research in relation to an ongoing site conservation programme where the variables of stone type and condition, lime and aggregate types and ratios and application procedure are already established. Testing should be carried out on stones, mortars and stone-mortar interface samples. Tests which should be of interest include porosity/water absorption by immersion and permeability (normally assessed indirectly by capillary rise of watering samples). Hydraulic lime binders should be included in any test programme. The movement of salt solutions through the stone-mortar interface should also be examined.

The benefits of ageing lime putty, increased plasticity and improved working properties are often cited, but are not entirely understood. Further research might identify the optimal storage time and conditions that would be practical, as well as the reactions in the lime putty responsible for the changes. Samples of mortar from fresh and aged putties could be prepared and tested for characteristics such as strength, shrinkage, plasticity and rate of carbonation.

Shelter coating

The use of other forms of lime than high calcium lime putty for shelter coating has not been tested by experience in the field or by laboratory-based experimentation. Hydraulic lime-based shelter coats have not been used to date, but field trials to determine whether or not they would be appropriate in certain circumstances would be of interest.

The role of casein, added in either skimmed milk or powder form, and its effect on the performance of lime shelter coats, has not been thoroughly explored. Investigation into how milk/casein modifies the lime shelter coat would be useful, including how it affects permeability of the coating and adhesion/bonding to the limestone.

Shelter coats are believed to protect limestone surfaces, partly by absorbing damaging salts and being converted to calcium sulphate in place of the stone. These intended functions have not been methodically investigated and verified. In order to effectively prevent surface conversion of carbonates to sulphates, shelter coats would need to act as a barrier to moisture and gases. The subject of protective coatings for weathered limestone is of some interest and comparative testing of shelter coats with other, mainly silicate-based, coatings may be useful.

ANNEX A. LIME TREATMENTS RECORD SHEET

Site:
Date(s) of stonework:
Date of inspection:
Inspected by:

A   record of work
1   Dates of conservation work:
2   Conservator/contractor:
3   Architect:
4   Records (include extent of written detail and recording, such as photography and verbal communication with architect/conservator)
5   Purpose of treatments if stated (consolidation, visual, etc)
6   Reasons for use of lime method (rather than other options)
7   Conservation work   Yes   No
       Cleaning
       Lime water
       Mortar repairs
       Repointing
       Shelter coat
8   Conditions of work (time of year, any protective measures)
9   Limestone type(s)
10   General notes on original condition of stone
11   Cost of project

B **Condition survey**
1 *Cleaning* (in preparation for lime treatment)
1.1 Reference area of treatment
1.2 General description of treatment area
1.3 Method (dry, wet, brushing, air abrasive, poultice, biocide)

2 *Mortar repairs*
2.1 Reference area of treatment
2.2 Description of treatment area
2.3 Reason for application (protective, aesthetic, etc)
2.4 Method of preparing surface
2.5 Method of application (any variations, details)
2.6 Material: mortar mix (include any details regarding preparation, condition and type of lime and aggregates) Present condition/appearance
2.7 Photographic reference
2.8 Method of examination (note use of binoculars, hand lens, etc)
2.9 Condition of repair (sound or deteriorating, describe appearance, any change in colour)
2.10 Summary of effectiveness of repairs/fills (include present appearance compared to that at time of application; evaluation of performance in terms of protecting stone or achieving the original aim of the treatment). Note effects on limestone, any limestone decay★

3 *Repointing*
3.1 Reference area of treatment
3.2 Description of treatment
3.3 Reason for use
3.4 Method of preparing joints
3.5 Method of filling joints
3.6 Material: mortar mix
3.7 Appearance/effect at time of application Present condition/appearance
3.8 Photographic reference
3.9 Method of examination
3.10 Condition of repointing (also note any stone decay)★
3.11 Effectiveness of treatment (in terms of protection, original aims)

4 *Shelter coat*
4.1 Reference area of treatment
4.2 Description of treatment area
4.3 Reason for application
4.4 Method of preparing the stone
4.5 Method of application (including number of applications)
4.6 Material: shelter coat mix (including any details of type/ condition of lime putty, and stone dusts)
4.7 Effect/appearance at time of application (reference existing photographs and documentation) Present condition/appearance
4.8 Photographic reference
4.9 Method of examination
4.10 Condition of shelter coated area of stone★
4.11 Summary of effectiveness of treatment (including evaluation of performance)

5 *Lime water*
5.1 Reference area of treatment
5.2 Description of treatment area (specify if done in conjunction with other lime treatments)
5.3 Reason for application
5.4 Method of surface preparation
5.5 Method of application (including number of applications)
5.6 Material: method of preparing lime water
5.7 Effect/appearance at time of application (ref existing documents/photographs) Present condition/appearance
5.8 Photographic evidence
5.9 Method of examination
5.10 Condition of surface★
5.11 Summary of effectiveness of treatment

6 *General assessment*
6.1 Was the treatment successful in achieving aims (of protecting stone, visual improvement, etc)?
6.2 Durability of treatment?
6.3 Generally was the treatment worth doing in terms of the original purpose?

★ Condition of limestone surfaces: describe any decay (e.g. powdery, friable, spalling, blistering, cracking, crazing)

Research team

This survey has identified the crucial role of technique and application method in the successful treatment of limestone. Any scientific investigation into the lime-based methods examined in this paper must involve the relevant scientists and experienced conservators if practical and reliable results are to be obtained.

ENDNOTES

1 The Preamble was written by Nick Durnan, a consultant stone conservator, who has extensive experience of the lime method. He provided much practical information for this report.
2 The cathedral architects Caroe and Martin, now renamed Caroe and Partners, inspect half of all the figure sculptures at Wells Cathedral in a six-year cycle. Much helpful information was also obtained from Jerry Sampson, cathedral archaeologist.
3 Records for the lime treatment trials at Malmesbury and 'before' treatment photographs of the five trial stones are held by English Heritage and include the work instructions and specification by John Ashurst, then of the Research and Technical Advisory Service, Directorate of Ancient Monuments and Historic Buildings (DAMHB), Department of the Environment. Stone conservator Richard Marsh also provided a record of the conservation work.
4 Professor John Ashurst was Director of Historic Building and Site Services (HBSS) in the Department (now School) of Conservation Sciences at Bournemouth University, which was commissioned by English Heritage to carry out this research. Professor Ashurst is now one of the principals of Ingram Consultancy, Gomshall, Surrey, UK.
5 Ammonium carbonate has long been a constituent of cleaning formulations used for removing sulfation soiling from wall paintings. The chemical reacts with calcium sulphate on the soiled surface to form calcium carbonate and ammonium sulphate, the latter being rinsed off (Durnan 1991b). A solution of 2.5 to 5% ammonium carbonate in water is prepared and a poultice produced by the addition of a variety of materials, including cellulose (typically carboxymethyl cellulose), clays or paper pulp. Other additives may be used, such as surfactants to increase wetting and reduce surface tension.
6 The solubility of lime in water increases with the addition of a few compounds, including glucose, and further research might look at the feasibility of including these compounds to promote deposition (Boynton 1980, 206). The use of glucose would, however, encourage organic growth, another adverse factor to consider.

BIBLIOGRAPHY

Alexander H and Sampson J, 1986 *Glastonbury Abbey, Lady Chapel, Conservation of the North and South Portals*, unpublished report.

Ashurst J, 1979 *Old Gorhambury: Specification for Repair and Surface Treatment*, Research and Technical Advisory Services, Department of Ancient Monuments and Historic Buildings, Department of the Environment, August 1979, unpublished report for English Heritage.

Ashurst J, 1983 The cleaning and treatment of limestone by the lime method. Part 1, in *Monumentum*, Autumn, 233–252.

Ashurst J, 1990 Surface treatments: Part 2, The cleaning and treatment of limestone by the lime method, in *Conservation of Building and Decorative Stone*, vol 2, (eds) Ashurst J and Dimes F, London, Butterworth Heinemann, 169–183.

Ashurst N, 1994 *Cleaning Historic Buildings*, 2 vols, London, Donhead.

Ashurst J and Ashurst N, 1988 *Stone Masonry*, volume 1, Practical Building Conservation, English Heritage Technical Handbook Series, Aldershot, Gower Technical Press.

Beadman K and Scarrow J, 1998 Laser cleaning Lincoln Cathedral's Romanesque frieze, in *Journal of Architectural Conservation*, 4(2), 39–53.

Boynton R S, 1980 *Chemistry and Technology of Lime and Limestone*, New York, John Wiley & Sons Inc.

British Standards Institution, 2000 *European Standard EN 459: 2000 Building Lime - Part 1: Definitions, specifications and conformity criteria*, London, Building Standards Institution.

Cooper M (ed), 1998 *Laser Cleaning in Conservation. An Introduction*, Oxford, Butterworth Heinemann.

Cowper A D, 1927 *Lime and Lime Mortars*, Building Research Special Report **9**, London, Building Research.

Durnan N, 1990 *Rochester Cathedral: The Cloisters, April 1990*, unpublished report.

Durnan N, 1991a *Rochester Cathedral: The Cloisters, February 1991*, unpublished report.

Durnan N, 1991b Ammonium carbonate cleaning of Caen stone, in *Conservation News*, **44**, 18–20.

Durnan N, 1991c *Exeter Cathedral: West Front, Conservator's Report*, unpublished report.

Durnan N, 1992 *Rochester Cathedral: The West Front, December 1992*, unpublished report.

Durnan N, nd *Canterbury Cathedral Romanesque Quire Arcade, Conservation Work 1986–1988*, unpublished report.

Holborow W A, 1992 *The Revival of Lime in Building Conservation*, unpublished MA thesis, Institute for Advanced Architectural Studies, University of York.

Marsh R, 1989 *Report, Schedule and Specification for the Tower's Repair and Conservation*, internal report for Basingstoke and Dean Borough Council.

Odgers D, 1987 *A Report on the Repair and Conservation of the Three Faces of the Chichester City Cross*, internal report for English Heritage.

Odgers D, 1993 Conservation of the West Front, Bath Abbey, in *Conservation News*, **51**, 13–14.

Sampson J, 1992 Glastonbury Abbey: The Lady Chapel doors and their dating in light of the 1986 conservation programme, in *The Conservator as Art Historian, Papers given at a UKIC Wall Paintings section conference, 20 June 1992, Abingdon, Oxfordshire*, (eds) Hulbert A, Marsden J and Todd V, London, United Kingdom Institute for Conservation of Historic and Artistic Works, 3–6.

Shekede L, 1998 The effects of laser radiation on polychromed surfaces, in *The Analysis of Pigments and Plasters: Post-Prints of a Day Conference of the Wall Paintings section of the United Kingdom Institute for the Conservation of Historic and Artistic Works on 22 February 1997*, London, UKIC, 24.

Sinclair E, 1992 Exeter Cathedral: exterior polychromy, in *The Conservator as Art Historian, Papers given at a UKIC Wall Paintings section conference, 20 June 1992, Abingdon, Oxfordshire*, (eds) Hulbert A, Marsden J and Todd V, London, United Kingdom Institute for the Conservation of Historic and Artistic Works, 7–14.

Teutonico J M, McCaig I, Burns C and Ashurst J, 1994 The Smeaton project: factors affecting the properties of lime-based mortars, in *Bulletin of the Association for Preservation Technology (APT)*, **25**: 3–4.

Woolfitt C and Durnan N, 1995 (revised 1996) *Lime Method Evaluation: A Survey of Sites*, Historic Building and Site Services, Bournemouth, Bournemouth University.

Zacharopoulou G, 1994 Bibliography of lime research, *Lime News*, **2**: 2, 42–60.

FURTHER READING

Ashurst J and Clarke B, 1972 *Stone Preservation Experiments*, Building Research Station, London, Department of the Environment.

Durnan N, 1991 *Rochester Cathedral: The West Front, November 1991*, unpublished report.

Honeybourne D B, 1990 Weathering and decay of masonry, in *Conservation of Building and Decorative Stone*, vol 1, (eds) Ashurst J and Dimes F, London, Butterworth Heinemann, 153–78.

Peterson S, 1982 Lime water consolidation, in *Mortars, Cements and Grouts used in the Conservation of Historic Buildings, Proceedings of ICCROM Symposium, Rome, November 1991*, Rome, ICCROM, 53–62.

Price C A, 1984 The consolidation of limestone using a lime poultice and limewater, in *Adhesives and Consolidants*, London, IIC, 60–162.

Price C A, 1996 *Stone Conservation: An Overview of Current Research*, Santa Monica, CA, Getty Conservation Institute, 1–11.

Price C A and Ross K D, 1984 The cleaning and treatment of limestone by the lime method Part II. A technical appraisal of stone conservation techniques employed at Wells Cathedral, in *Monumentum*, 301–312.

Price C, Ross K and White G, 1988 A further appraisal of the lime technique for limestone consolidation, using a radioactive tracer, in *Studies in Conservation*, **33**, 178–186.

Schaffer R J, 1932 *The Weathering of Natural Building Stones*, Building Research Establishment Special Report No 18, Watford, Building Research Establishment.

Skoulikidis T and Papakonstantinou P, 1992 Stone cleaning by the inversion of gypsum back into calcium carbonate, in *Stone Cleaning and the Nature, Soiling and Decay Mechanisms of Stone*, (ed) Webster R G M, London, Donhead Publishing, 155–9.

Tiano P, 1995 Stone reinforcement by calcite crystal precipitation induced by organic matrix molecules, in *Studies in Conservation*, **40**, 171–176

ACKNOWLEDGEMENTS

The significant contribution of Nick Durnan, stone conservator and author of the preamble, to the survey work and to the draft report upon which this chapter is based, is gratefully acknowledged. The author would also like to thank Professor John Ashurst, Jerry Sampson, Hebe Alexander, Richard Marsh and David Odgers for useful discussion on all aspects of the lime method.

AUTHOR BIOGRAPHY

Catherine Woolfitt trained in classical archaeology and art conservation, specializing in artefacts. She carried out the lime method survey while working with Historic Building and Site Services of the School of Conservation Sciences at Bournemouth University. She worked as a consultant in the conservation of historic buildings and sites with Hutton + Rostron's Resurgam™ conservation consultancy before becoming co-Director of Ingram Consultancy.

Nicholas Durnan is a sculpture conservator and stonecarver, who has worked on the repair and conservation of historic stonework on English cathedrals for over 25 years. He has been involved at Salisbury for the last eight years, where he is consultant conservator. He has also been involved with major architectural and sculptural conservation projects at Canterbury, Exeter, Rochester and Wells Cathedrals, and Westminster Abbey.

An investigation of sacrificial graffiti barriers for historic masonry

Nicola Ashurst
Adriel Consultancy, 22 Hockley, Nottingham NG1 1FP, UK

Sasha Chapman and Susan MacDonald
English Heritage, 23 Savile Row, London W1S 2ET, UK

Roy Butlin and Matthew Murry
Building Research Establishment, Garston, Watford WD2 7JR, UK

Abstract

The removal of graffiti from external surfaces is a large industry in the UK, with historic building surfaces being as much at risk as more modern surfaces. The removal of these unwanted markings is compounded by the staining and other damaging characteristics of many of the marker materials. In order to investigate the nature and effectiveness of sacrificial graffiti barriers, from 1993–6 English Heritage commissioned the Building Research Establishment (BRE) to carry out research into the nature and effects of sacrificial graffiti barriers on historic masonry surfaces. The BRE was asked to develop and evaluate methodologies for assessing the suitability and effectiveness of proprietary sacrificial graffiti barriers for use on historic masonry, with particular attention to assessing the effects on the weathering characteristics of barrier-treated stone, brick and terracotta. This paper is a summary of the research and its key findings, and provides historic building practitioners with advice on the value and use of these materials.

Key words

Sacrificial barriers, graffiti, damage, reversibility, protection, sandstone, limestone, terracotta, brick

INTRODUCTION

Context

The removal of graffiti from external surfaces is a multimillion pound industry in the UK, with historic building surfaces being as much at risk as more modern surfaces. The cost and annoyance of removing these unwanted markings are compounded by the tenacious staining and damaging characteristics of many of the marker materials and the damage that often results.

Current techniques to remove graffiti involve using solvents, caustic solutions, detergents and abrasives. These work by dissolving, bleaching, washing or abrading graffiti respectively. As the surfaces of historic buildings are often composed of porous and delicate materials, the removal of graffiti from these structures can damage the underlying surface. Residues of graffiti must be accepted as a necessary consequence of minimizing damage to the masonry by normal removal procedures.

Graffiti barriers

Impervious barriers have been used for many years to protect areas of buildings vulnerable to graffiti attack. The inks and dyes from the graffiti stay on top of the barrier and are prevented from staining the building. These graffiti barriers also provide a surface from which graffiti can be removed more easily. Many have been found to cause problems on porous surfaces because they either slow down or prevent water evaporation, increase the damaging effects of soluble salt crystallization and cause a marked colour or appearance change to the substrate. This has meant that their use on historic buildings has rarely been regarded as acceptable.

Sacrificial graffiti barriers

Sacrificial graffiti barriers were developed as an attempt to solve the problems of the impervious barriers. Sacrificial graffiti barriers are removable barriers based on organic compounds such as waxes or carbohydrates which are removed by melting using a high-pressure hot water spray or dissolving with a graffiti-removal product. The graffiti on these barriers is loosened and washed away with the coating. Sacrificial barriers have other advantages over normal graffiti barriers which make them of interest for use on historic buildings. These include a reduced effect on surface appearance, ease of removability and permeability to moisture.

The protection of areas of historic masonry from extremely high levels of graffiti attack by a sacrificial graffiti barrier may, however, introduce a new set of problems and considerations of effectiveness, masonry durability and changes in appearance.

The research programme

In order to investigate the nature and effectiveness of sacrificial graffiti barriers, from 1993–6 English Heritage commissioned the Building Research Establishment (BRE) to carry out research into the nature and effects of sacrificial graffiti barriers on historic masonry surfaces. The BRE was asked to develop and evaluate methodologies for assessing the suitability and effectiveness of proprietary sacrificial graffiti barriers for use on historic masonry, with particular attention to assessing the effects

---

* Correspondence: Bill Martin, Building Conservation & Research Team, English Heritage, 23 Savile Row, London W1S 2ET, UK

on the weathering characteristics of barrier-treated stone, brick and terracotta.

This paper is a summary of the research and its key findings, and provides historic building practitioners with advice on the value and use of these materials. (Further work required to investigate these coatings and associated graffiti removal processes is also described.)

THE CHARACTERISTICS OF GRAFFITI BARRIERS

Desirable characteristics of sacrificial graffiti barriers

At the beginning of the research, the desirable characteristics of sacrificial barriers were defined as:

- appearance: a sacrificial graffiti barrier should cause no visible change in surface appearance.
- reversibility: the barrier should be fully reversible, and traces of the barrier should not be left on nor absorbed into the surface of the masonry.
- effect on masonry durability: the durability of the treated masonry surface should not be reduced.
- efficiency of graffiti removal: the removal process should completely remove all graffiti.

The tests selected for the analysis of the barriers and their effects were aimed at investigating these characteristics. Procedures used and the resultant findings are presented below. It was expected that sacrificial graffiti barriers would behave differently on different surfaces. The following substrates were selected for inclusion in the tests in a variety of combinations:

- brick
- weathered brick
- terracotta
- weathered terracotta
- mortar
- sandstone
- weathered sandstone
- limestone
- weathered limestone.

Selection of sacrificial graffiti barriers for testing

When the research began only a limited number of sacrificial graffiti barriers were available. The coatings were categorized according to generic type:

- emulsions: acrylics or microcrystalline waxes suspended in water
- solutions: consisting of the coating dissolved in a volatile organic solvent
- gels and other suspensions: a broad category consisting of water mixed with hydrophilic polymers such as polysaccharides. These can be synthetic or naturally produced.

The range of sacrificial graffiti barriers available throughout the research period was constantly changing. Only products which appeared to be consistently available in the UK were selected. Three barriers were initially chosen. Two were from the first category (emulsions). Within this group, one represented an acrylic polymer, the other a wax. The third was from the second category (solutions), consisting of a wax dissolved in white spirit. The products [1] were given the following labels:

SGB1 – Microcrystalline wax emulsion in water
SGB2 – Microcrystalline wax dissolved in white spirit (later eliminated from the tests)
SGB3 – Acrylic polymer emulsion in water

Most of the tests concentrate on SGB1 and SGB3 as this provided a comparison between different manufacturers' commercial products. The time frame and scope of this research project did not enable all tests to be applied to all products.

Initial selection of the range of products was made on the basis of visual effect, ease of application and general appearance. The final selection of sacrificial graffiti barriers was made by testing three barriers simultaneously on a sandstone wall. Graffiti were added to these strips of coating and water pressures of approximately 20, 40, 60 and 80 Bar (300, 600, 900 and 1200 psi) were used to remove them. The two barriers finally selected (SGB1 and SGB3) were equally effective at graffiti protection in this simple test.

Appearance

'Acceptable appearance' was deemed to mean that the difference in appearance of surfaces before and after treatment with graffiti barriers should be minimal. Ideally, there should be no change at all.

The visual effect of the selected sacrificial graffiti barriers on two types of limestone (Portland and Monks Park) and one sandstone (White Mansfield Dolomitic) was evaluated. In addition, samples of Portland limestone treated with four different quantities of sacrificial graffiti barrier were assessed. All assessments were made by using colour and gloss meters before and after treatment.

Colour change from the application of two coats of barrier was measured for all sample surface types using a colour spectrophotometer. For each situation, 32 colour and gloss measurements were taken over a 15 mm x 15 mm (approximately $\frac{1}{2}$ inch) grid using an X-rite 918 colourmeter and a Pico-glossmaster 500 at an angle of 60°. Colour and gloss measurements were successful in identifying visually imperceptible changes in surface appearance between coating types, substrate types and numbers of applications.

Colour change
It was found that all barriers tested (SGB1, SGB2 and SGB3) caused small colour changes on all material types when they were applied. On the darker stones, the colour change produced by one of the barriers was insignificant,

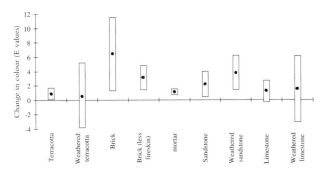

Figure 1. Change in colour of different masonry types coated with SGB1. The upper and lower limits represent the standard deviation of colour from the untreated substrates.

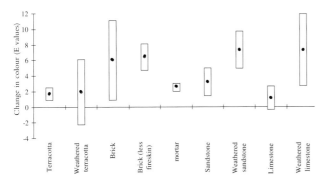

Figure 2. Change in colour of different masonry types coated with SGB3. The upper and lower limits represent the standard deviation of colour from the untreated substrates.

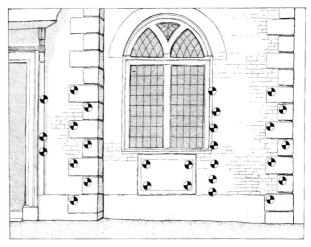

Figure 4. The Old Free School, Watford; positions of colour and gloss measurements taken.

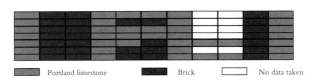

Figure 5. The Old Free School, Watford; positions of colour measurements and surface type.

while for another there was a slightly milky appearance visible to the naked eye. While the colour change was generally insignificant on non-weathered surfaces, it was more apparent on weathered surfaces. This may occur for two reasons. Firstly, a rough surface texture and any lichens or growths on the weathered surfaces can retain more of the barrier. Secondly, the colour of the existing surface soiling and particulates can be enhanced by the barrier and appear as a general darkening. Each sacrificial graffiti barrier selected had a different effect on visual appearance, post-application (Figs 1 and 2).

Gloss change

It was found that gloss increased as the quantity of applied barrier increased. At a more detailed level, the barriers tested were found to behave differently. SGB1 caused a small increase in gloss on terracotta, and a very small increase in gloss on weathered limestone. No residual gloss was found after removal. SGB3 caused an increase in gloss on all samples except brick. After repeated retreatment and removal, an increase of gloss value was recorded for sandstone and weathered limestone. For both of the barriers (SGB1 and SGB3), terracotta surfaces exhibited the strongest gloss change (see Fig 3).

Colour and gloss measurements were also taken from an historic building constructed of brickwork and Portland limestone (Figs 4 and 5). Graffiti was removed from the brick and stone surfaces before the application of the sacrificial graffiti barriers. Colour and gloss values were measured after graffiti removal and then after the application of the sacrificial graffiti barriers, and again at intervals of two weeks and two months (Figs 6 and 7). Significant colour changes were identified and attributed to the initial graffiti removal where general pressure washing was used. No visible changes in appearance occurred due to the subsequent application of the sacrificial graffiti barriers. However, colour changes did occur due to the development of efflorescence.

Reversibility and depth of penetration

It is always important to determine the structure of coatings on historic surfaces. The coating thickness, and the way it covers natural pores in the masonry surface, can alter the weathering characteristics of the masonry. It is also highly desirable to be able to completely remove sacrificial graffiti barriers. A section of the research programme was dedicated to the investigation of these aspects.

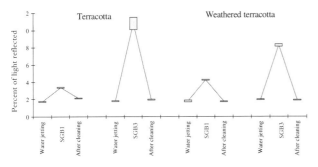

Figure 3. Change in gloss values of different surfaces treated with SGB1 and SGB3. The blocks represent the standard deviations of the gloss data.

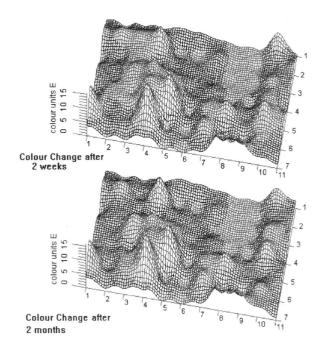

Figure 6. The Old Free School, Watford. Colour change after cleaning and treatment.

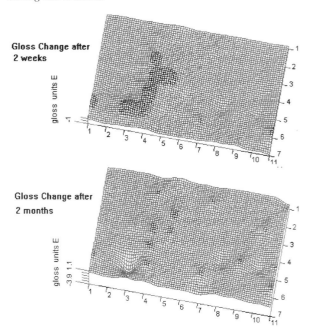

Figure 7. The Old Free School, Watford. Gloss change after cleaning and treatment.

Methods of analysis

The following analytical procedures were used to inspect treated masonry.

- SEM (Scanning Electron Microscopy). The barriers were sometimes visible using SEM from a photograph-style image (Fig 8).
- XRD (X-Ray Diffraction). Detection of barriers was possible by X-ray mapping of carbon.
- FTIR (Fourier Transform Infrared Spectroscopy). This technique was used to detect and quantify the depth of barrier penetration. Drillings were taken at 0.5 mm increments from treated and untreated samples of Portland limestone and weathered White

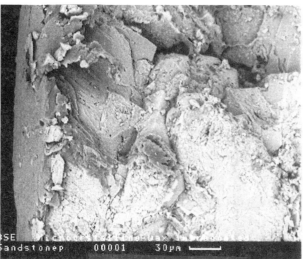

Figure 8. An SEM photomicrograph of Portland limestone (top) and White Mansfield sandstone (above) treated with SGB2 (BRE for English Heritage).

Mansfield sandstone. The powder was then analysed by diffuse reflectance FTIR spectroscopy. The relative concentration of graffiti barrier was determined from the infrared absorption spectra of the wax present in the barrier. By comparing the cleaned to uncleaned FTIR spectra, it was evident that the barrier was not totally removed from the stone. It was estimated that 80% of the first barrier tested stayed in the top 0.5 mm on Portland limestone and 98% stayed in the top 0.5 mm on the weathered sandstone. All residues remained in the top 0–0.5 mm layer. No barrier was detected at greater than 1 mm depth into either surface (Fig 9).

- Chemical staining and visual inspection. Samples of the sacrificial graffiti barriers were stained using iodine vapour. Visual observation and photographic analysis were used to determine barrier residues and any other surface characteristics (Figs 10 and 11).

As both SGB1 and SGB3 were in the form of an emulsion, ie many small solid particles suspended in

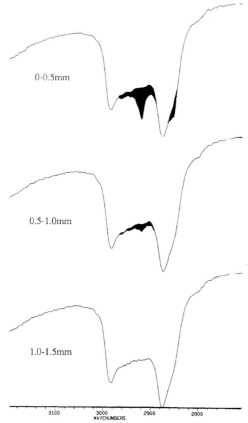

Figure 9. The FTIR spectra of drillings taken at 0.5mm increments in Portland limestone treated with SGB1. The black area is the contribution from the barrier.

Figure 11. SGB3 stained black with iodine on a limestone surface viewed in cross-section. Field of view 2x3 mm (BRE for English Heritage). See colour plate 17.

surfaces with larger holes and pores, the original pore structure was not significantly affected. The pore structure of the surfaces to which they were applied was probably the most significant factor in determining whether a continuous or discontinuous film would be formed by the sacrificial graffiti barrier (Fig 12).

Reversibility (degree of removability)

It is likely that water jetting temperature, water rinsing pressure, cleaning technique, application technique and masonry surface characteristics will affect the degree of reversibility of a sacrificial graffiti barrier. The analytical procedures confirmed that 100% removal of the sacrificial graffiti barriers tested could rarely be achieved.

water, they tended to be viscous and were not absorbed deeply by capillary action. In the stones tested, the barriers did not travel further than 1 mm into masonry surfaces but tended to stay on the surface of masonry, only entering as far as pore spaces would allow (Fig 11).

In summary, the analytical techniques deployed showed that no barrier penetrated further than about 0.5 mm into the stone, except where there were large pores or holes. The barriers tended to coat the surface and dip into pore spaces, rather than to be absorbed directly into the substrate. Consequently, on surfaces where pores were generally small, a continuous layer 0.05–0.1 mm thick was formed, sometimes filling pores and holes. On

Figure 10. Photograph of SGB1 applied to Monks Park limestone, clearly showing the depth of penetration and the structure of the barrier over the surface. (BRE for English Heritage) See Colour Plate 16.

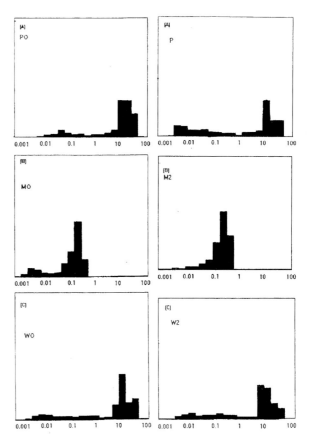

Figure 12. The pore size distribution in microns of untreated (left) and barrier treated (right) stone cubes. (A) Portland limestone (B) Monks Park limestone (C) White Mansfield sandstone.

A. Monks Park limestone

B. Portland limestone

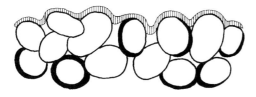

C. White Mansfield sandstone

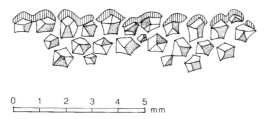

0 1 2 3 4 5 mm

Figure 13. Sketches of the views seen through the microscope of SGB1 applied to (a) Monks Park limestone (b) Portland limestone (middle) and (c) White Mansfield sandstone.

After repeated removal and reapplication of barriers, residual coatings start to remain on masonry surfaces. By using higher water pressures than those tested it may be possible to remove these traces, but this risks damage to the masonry. FTIR could be used to observe barrier build-up as the result of multiple use, and permeability measurements could determine permeability changes if these factors were to be investigated further.

Sacrificial graffiti barriers are marketed as being suitable for repeated use on the same substrate surface. However, the effect of repeated graffiti application and removal processes can build up concentrations of the barrier, or other materials such as fine stone fragments, for example, in the surface of the masonry. This blocking effect could be extremely damaging as it would affect the surface permeability of the masonry.

Effect on masonry durability

To be suitable for use on porous, historic masonry, sacrificial graffiti barriers should have little or no effect upon the durability of masonry. In order to test and attempt to predict the durability of barrier-treated materials and buildings, several weathering characteristics of a range of masonry types were assessed to establish the changes a barrier might cause. Characteristics tested were:

- porosity
- permeability
- freeze/thaw
- salt crystallization

The materials tested were:

- Monks Park limestone
- Portland limestone
- White Mansfield sandstone

Long-term weathering tests
For the most definitive results, barriers would need to be tested on actual buildings, rather than isolated blocks of stone. These tests will be necessary to assess the long-term weathering characteristics of stone treated with sacrificial graffiti barriers.

Porosity/pore size distribution
Porosity is a measure of the volume of pores within stone or brick. Porosimetry provides an indication of pore size and distribution. Since pore size is one of the factors influencing the durability of stone, it was important to investigate how this characteristic might be affected by the absorption of a sacrificial graffiti barrier. In particular, because salt crystallization causes damage by forming salt crystals within small pore spaces (less than 1µm diameter), the research sought to determine whether the number of pores in this size range was decreased by the sacrificial graffiti barrier.

The test involved mercury porosimetry, where mercury is forced into the stone surface under pressure. The change in pressure and volume correlates to the stone pore sizes on the surface of the barrier-coated masonry.

The results showed some reduction in pore size at the 60µm size after both one and two treatments of a selected sacrificial graffiti barrier (see Fig 13), although there was no noticeable difference between the application of one or two coats (Fig 14). However, the barrier did not affect the percentage of smaller-sized pores for all three stone types tested. In contrast, changes in porosity were found due to the filling-up of the larger pore spaces (a reduction in porosity of 75% in White Mansfield, 35% in Monks Park and 20% in Portland).

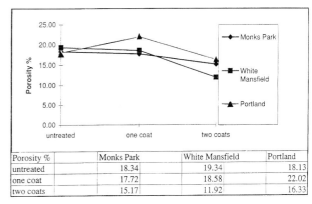

Porosity %	Monks Park	White Mansfield	Portland
untreated	18.34	19.34	18.13
one coat	17.72	18.58	22.02
two coats	15.17	11.92	16.33

Figure 14. The porosity of stone types with multiple graffiti barrier treatments.

Table 1. Comparison of water vapour permeability for stone substrates with and without SGB 1, in g/m² per day of water.

stone type	untreated mean	standard deviation	treated mean	standard deviation	difference and % drop		T-test 95%
Portland limestone	61	24	42	12	19	31%	fail
Monks Park limestone	66	7	48	5.4	18	27%	pass
White Mansfield sandstone	86	2.7	73	6.2	13	15%	pass

Water vapour permeability changes

Changes in moisture permeability of a building material surface affect the rate of water evaporation. In turn, this can affect soluble salt crystallization. A barrier coating over the surface would normally be expected to reduce permeability. This effect is likely to encourage damaging crystallization further inside the material.

The first series of tests conducted were based upon standard water vapour and permeability tests for natural stone (BSI 3177:1959 [1995]). They involved measuring the weight change as moisture passed through the barrier-treated masonry into a deliquescent (water-absorbing) material sealed underneath the coatings. Permeability tests were undertaken on samples of Portland limestone, White Mansfield sandstone and Monks Park limestone. All barriers caused a decrease in permeability. White Mansfield sandstone showed the smallest change in permeability when treated with SGB1 compared with Portland limestone and Monks Park limestone (Table 1).

The reductions in permeability caused by the application of sacrificial graffiti barriers must be considered significant. Only a very small change may be required to alter moisture movement in masonry surfaces in such a way that crystallization damage is increased.

In a second series of tests, permeability before and after treatment with barrier on the same sample was carried out to determine more exact permeability changes. This time, daily permeability rates were measured for each sample before and after treatments through a test based on BS 3177: 1959 [1995]. This enabled the percentage change in permeability of the substrate due to the sacrificial graffiti barrier application to be calculated. Weathered and un-weathered limestone and sandstone were subjected to three further cycles of barrier application and removal, to test the re-application of barriers. The key findings are given below (Figs 15 and 16).

For the first sacrificial graffiti barrier tested (SGB1), a decrease in permeability was recorded for all materials on the first coating, the highest values being a reduction of 40% on weathered terracotta, 10% on brick, and about 10% for limestone and sandstone. Three subsequent retreatment processes (application and removal) did not cause any progressive decrease in permeability, and, for weathered limestone and weathered sandstone, the permeability appeared to increase on further treatments. This is believed to be due to the repetitive use of high-pressure water jetting which was shown to increase the permeability of sandstone and limestone by removing weathered surface layers.

For the second sacrificial graffiti barrier under assessment (SGB3), a decrease in permeability was recorded for all materials, the highest values being a reduction of 40% for terracotta, 30% for brick, about 10% for limestone and 60% for sandstone. Substantial further reductions also occurred on retreatment of limestone and sandstone, increasing to over 60%.

On final cleaning of retreated limestone and sandstone samples, there was a dramatic increase in permeability, back to original levels in many instances.

Freeze/thaw testing

It is well known that water under a surface barrier that limits permeability can freeze, causing damage to the substrate. Treated and untreated samples of masonry underwent freeze/thaw trials based on prEN 12371 freeze/thaw resistance test based on the draft available in 1997. Treated and untreated limestone samples and mortar samples 'crumbled' after 20 cycles. The experiment was stopped after 50 cycles with no further damage having taken place to other samples.

The test showed no relationship between barrier treatment and failure of the samples, indicating that the barriers had no effect on frost resistance. The pattern of breakdown of the masonry samples did not follow the contour of the treatment, confirming that the barriers had not affected the path of frost decay. However, it must be remembered that simple freeze-thaw tests are not representative of real frost action upon a wall. They do not simulate the direction of freezing or include for ionic salts dissolved in the water which can alter the effect of frost damage.

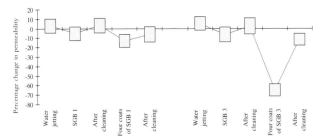

Figure 15. Permeability changes due to water cleaning, graffiti barriers and removal of barriers on limestone.

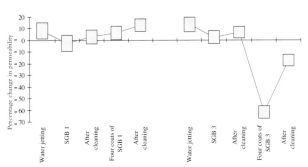

Figure 16. Permeability changes due to water cleaning, graffiti barriers and removal of barriers on weathered limestone.

Figure 17. Samples of Portland limestone treated with SGB2 (left) SGB1 (middle) and untreated (right) being subjected to the crystallization test (BRE for English Heritage). See Colour Plate 18.

Figure 18. Efflorescence observed on samples of Portland limestone treated with SGB2 (left) SGB1 (middle) and untreated (right) after three days of the crystallization test (BRE for English Heritage). See Colour Plate 19.

Figure 19. The final condition of samples of Portland limestone treated with SGB2 (left), SGB1 (middle) and untreated (right) after two weeks of the crystallization test (BRE for English Heritage). See Colour Plate 20.

Soluble salt crystallization
A major part of masonry decay in the UK is due to soluble salt crystallization. The movement and concentration of the salts is affected by the water vapour permeability of the stone. If a barrier placed over a stone surface reduces the surface permeability, it will reduce water loss by evaporation, and could deepen crystallization effects, increasing damage to the substrate.

Standard tests to assess crystallization do not represent a good simulation of a real weathering situation. They are, however, the only accelerated way of forming any quantitative assessment of crystallization damage.

Capillary rise zone crystallization test

The mechanism of decay due to crystallization of masonry treated with sacrificial graffiti barrier was studied using a variation of an existing BRE test (Lewin 1982). The method involved standing 300 mm x 40 mm x 40 mm (12 in x 1.5 in x 1.5 in) columns of stone treated with two types of barrier in saturated sodium sulphate solution to mimic the decay caused by rising ground water salts. As the solution soaked into the stone and evaporated, different degrees of crystallization occurred along the vertical stone columns. Two types of stone were used, Portland limestone and White Mansfield sandstone, two samples of each were treated with two coats of barrier and one of each was left untreated. Four days after treatment, the samples were subjected to the test for two weeks at ambient room temperature and humidity.

The effects of the saturated solution were recorded on a daily basis. Different patterns of efflorescence, contour scaling, height of capillary rise, etc were observed in each material and with each barrier type (Figs 17–19).

The White Mansfield sandstone showed similar decay patterns with and without SGB1 treatments. Portland limestone, however, showed a marked change in crystallization decay between SGB1 treated and untreated samples. On treated Portland limestone, soluble salt crystallization took place further into the body of the stone.

Different stone types showed significantly different behaviour in salt crystallization when treated with SGB1. SGB1 applied to White Mansfield sandstone appeared to be 'shed' during crystallization by scaling of the stone surface, allowing further natural crystallization action similar to the untreated sample. This did not occur with the Portland limestone where SGB1 induced crystallization damage deep within the stone.

SGB2 applied to White Mansfield sandstone promoted scaling over large areas of the surface of the sample. Portland limestone did not sustain any visible decay of this sort, probably because SGB2 reduced the evaporation of moisture from the sample. When the Portland limestone sample was broken open large amounts of sodium sulphate were found deep within the stone. This would probably have caused major damage if the experiment had been continued.

Changes in the permeability and porosity of surfaces treated with sacrificial graffiti barriers, as well as durability in salt crystallization, can be assessed in the laboratory, but such results do not necessarily accurately predict field performance. The capillary rise crystallization test provided a good way of assessing the path of crystallization decay of barrier-treated surfaces. However, further study is needed, involving long-term assessment of stone samples placed on site, together with observations of barriers applied to real buildings.

Figure 20. The result of high-pressure hot water cleaning from sacrificial graffiti barrier treated limestone samples (SGB1, top, and SGB3, upper middle) and weathered limestone samples (SGB1, lower middle, and SGB3, bottom). Green spray paint and black felt marker pen were used as graffiti types. The control samples were not treated with a barrier and were cleaned with high pressure water only (BRE for English Heritage). See Colour Plate 21.

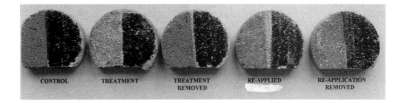
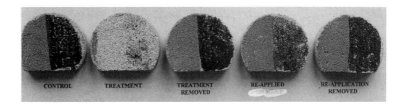

Assessment of durability by capillary rise crystallization tests showed that:

- the development of salt crystallization within each masonry type was affected by the presence of a sacrificial graffiti barrier on the surface.
- different sacrificial graffiti barriers promoted different depths of salt crystallization on the same substrate type.
- different stone types showed significantly different behaviour in salt crystallization when tested with a particular sacrificial graffiti barrier.

Based on these results it is clear that sacrificial graffiti barriers of the types tested should not be applied where soluble salts are an existing or potential problem.

Efficiency of graffiti removal

Investigation of graffiti removal efficiency began by establishing an optimum range for the quantity of sacrificial graffiti barrier applied. Results confirmed that application of increased amounts of SGB1 did not actually mean increased effectiveness in the removal of graffiti.

Limestone and sandstone samples were coated with SGB1 and SGB3 and defaced with green acrylic spray paint and felt marker pen. High-pressure hot water was used to remove the barriers and the graffiti (Figs 20 and 21).

This aspect of the test programme began to demonstrate the range of difficulties that can be experienced in removing different types of graffiti. Substrates with pore structures were particularly difficult to clean without damage. The graffiti type also affected the type of residual staining. The green spray paint tended to form patches on

Figure 21. Abrasive damage to stone as a result of increased pressure. 20 bar (300 psi), 40 bar (600 psi), 60 bar (900 psi), 80 bar (1200 psi) (BRE for English Heritage). See Colour Plate 22.

Table 2. Units of pressure used

psi	MPa	bar
300	2.07	20.7
600	4.14	41.4
900	6.21	62.1
1200	8.27	82.7
1500	10.34	103.4

top of the barriers, the felt pen tended to penetrate and stay under or within the barriers.

Residual staining was due to two causes:

- graffiti that had penetrated through the barrier and stained the masonry underneath. This graffiti could not be removed by increasing the pressure of the water. It was not removed when the barrier was removed. Marker pen ink containing pigment proved especially problematic. If a marker pen was pressed hard, the barrier could be penetrated by the ink.
- remains of graffiti left on top of the barrier, or staining the barrier itself. These residues could be removed by hot water rinsing. However the level of water pressure required to do this damaged the masonry.

There were marginal differences in the effectiveness of the two sacrificial graffiti barriers studied as measured by their ability to permit removal of the two graffiti types from the various substrates.

As an extension to the second series of tests, limestone and sandstone surfaces were tested after repeated applications (four in total) of graffiti in the same location. The retreatments appeared to reduce the barrier effectiveness.

THE EFFECT OF PRESSURE WATER ON HISTORIC MASONRY

High-pressure water is often used in sacrificial graffiti barrier removal and general building cleaning. It can be a very aggressive cleaning technique which tends to be underestimated in its ability to damage surfaces, especially those in a friable condition. The type and condition of masonry will therefore determine the pressure of water jetting which can be used for graffiti removal, even with

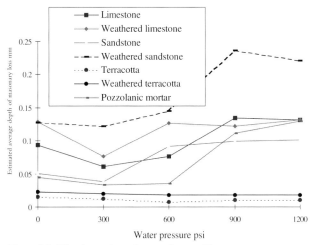

Figure 23. The change in surface roughness with water pressure on a range of masonry substrates.

newly cut, unweathered materials. Generally, weathered masonry on historic buildings will be less durable than newly-cut masonry. Table 2 sets out comparative units of pressure used in the research.

To confirm initial suspicions about the impact pressure water can have on masonry, three samples of Portland limestone, Monks Park limestone and White Mansfield sandstone, without graffiti barriers, were cleaned with high-pressure hot water at approximately 103.4 bar (1500 psi). This is the pressure usually recommended by the

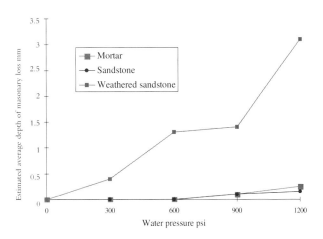

Figure 22. The loss of masonry surface at different water pressures on a range of masonry substrates.

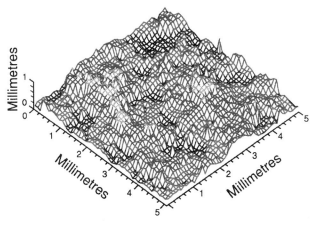

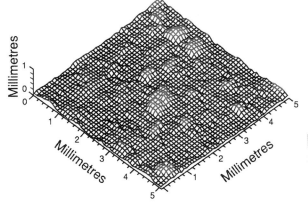

Figure 24. Surface profile comparison of Monks Park limestone, water-cleaned at approximately 100 bar (1500psi) (above) and uncleaned (below).

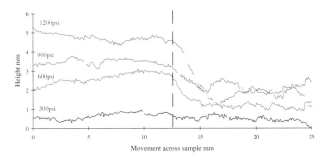

Figure 25. The surface roughness of weathered sandstone cleaned with high pressure hot water at different water pressures. Only the range 12.5-25mm (1/2–1 ins) was cleaned.

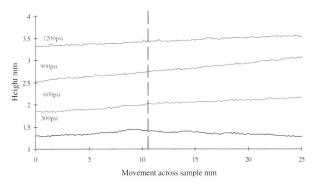

Figure 27. The surface roughness of a terracotta surface cleaned with high pressure hot water at different water pressures. Only the range 12.5–25mm (1/2–1 in) was cleaned.

manufacturers of sacrificial graffiti barriers for their removal. Half of each sample surface was left uncleaned as a control.

Two 5 mm x 5 mm (3/16 in x 3/16 in) regions were then selected on each sample, one from the cleaned area and the other from the control area. These areas were analysed by laser surface profilometer. The results showed that, due to water cleaning at 103.4 bar, there was major surface pitting of Monks Park limestone and significant surface removal from Portland limestone and White Mansfield sandstone.

To gain additional information regarding the effects of high-pressure hot water, four areas were identified for further study:

- surface damage
- change in permeability
- change in colour
- effects of temperature.

Surface damage

Figures 22 to 28 and Table 3 all relate to the investigation of surface damage. The aim of these trials was to establish the highest water pressure that could be used without causing surface damage, as measured by changes in surface roughness and masonry loss. The work was undertaken on untreated masonry samples. The spray profile of the pressure water system[2] produced a broad V shape from the nozzle. The use distance was about 100 mm from nozzle to substrate.

The induced surface roughness showed a characteristic relationship to water pressure.

- 0–20.7 bar (0–300 psi) resulted in a decrease in surface roughness compared to the roughness level before cleaning. This was probably due to the removal of loose dust, particles or masonry grains held together by organic growths or weathering products.
- 20.7–62.1 bar (300–900 psi) gave rise to an increase in surface roughness compared to the roughness level before cleaning. This was probably caused by the removal of loosely-bound grain particles within the masonry which gave a partially etched surface. There may or may not have been damage to the non-decayed masonry surface.
- at 62.1–82.7 bar (900–1200 psi), there was a levelling off of surface roughness, or even a slight decrease

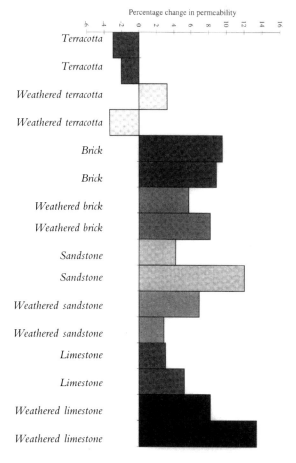

Figure 28. The change in permeability due to high pressure water cleaning on a range of masonry substrates.

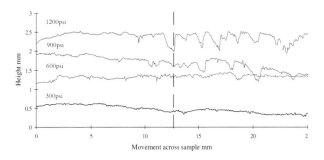

Figure 26 The surface roughness of pozzolanic mortar cleaned with high pressure hot water at different water pressures. Only the range 12.5–25mm (1/2 in–1 in) was cleaned.

Table 3. *The level of graffiti removal (visually assessed) from a range of masonry surfaces. It should be pointed out that 62.1 bar (900 psi) water jetting was used and that the effects would be different at higher or lower water pressures. Residual staining of graffiti penetrated through the barrier and stained the masonry underneath; this was not removed when the barrier was removed, and could not be eliminated with increased water pressure. Partially removed graffiti was also sometimes left on top of the barrier, forming patches. This was especially visible on the re-treated samples. This second type of residual graffiti could be reduced with higher water pressure.*

Water jetting alone	terracotta	weathered terracotta	brick	weathered brick	sandstone	weathered sandstone	limestone	weathered limestone
green spray paint	P	C	U	U	U	U	U	U
black felt pen	U	U	U	U	U	U	U	U
SGB3	terracotta	weathered terracotta	brick	weathered brick	sandstone	weathered sandstone	limestone	weathered limestone
green spray paint	C	C	U	U	P	U	C	U
black felt pen	C	C	C	C	S	S	S	U
SGB1	terracotta	weathered terracotta	brick	weathered brick	sandstone	weathered sandstone	limestone	weathered limestone
green spray paint	C	C	C	C	C	U	S	S
black felt pen	S	S	S	C	U	U	U	U
Re-treated SGB3	terracotta	weathered terracotta	brick	weathered brick	sandstone	weathered sandstone	limestone	weathered limestone
green spray paint					U	U	U	U
black felt pen					P	U	U	U
Re-treated SGB1	terracotta	weathered terracotta	brick	weathered brick	sandstone	weathered sandstone	limestone	weathered limestone
green spray paint					S	U	U	S
black felt pen					U	U	U	U

U: uncleaned; cleaning system failed to remove graffiti
P: cleaned patches
S: cleaning with residual staining
C: complete removal of graffiti

compared to the roughness level before cleaning, but at the expense of significant masonry loss.

Figure 21 shows the visual effects of the above tests on a piece of weathered sandstone. Even at 20.7 bar (300 psi), surface damage is visible and damage increases as the pressure increases.

The surface loss due to applied water pressure observed in the experiments indicated two possible trends:

- up to 62.1 bar (900 psi), the loss of masonry surface due to water pressure showed a roughly linear relationship. The weathered sandstone sustained significant surface loss, even though there was a slight decrease in surface roughness.
- over 62.1 bar (900 psi), the damage to mortar and weathered sandstone continued at a faster rate. The water pressure started to break down the structure of the masonry surface.

Based on these results, it was concluded that where pressure water equipment, of a type similar to that employed for the trials, is used, only water pressures below 20.7 bar (300 psi) should be used in cleaning and graffiti removal on most weathered and historic masonry; on more durable, modern materials, pressures below 62.1 bar (900 psi) may be used. However, the pressures indicated for the situations cited do not apply to all types of equipment and modes of use. In general, the pressures recommended by suppliers of chemical graffiti removal products are not suitable for historic masonry and need to be significantly reduced. The final pressure selected should always be based on site testing, taking into full account the nature and condition of the substrate and its joints.

On very friable surfaces consolidated with gypsum, organic growths or residual geological cementing, any physical force from water jetting may result in significant surface loss.

Changes in permeability due to high–pressure water cleaning

High pressure water may remove material from a masonry surface which, in turn, will alter its permeability. To determine the effects of pressure washing on permeability, samples were tested before and after high-pressure water cleaning using a customized permeability test based on BS 3177: 1959 [1995].

All brick, sandstone and limestone samples showed increased permeability after cleaning. On average, terracotta decreased in permeability (Fig 28).

Colour change

Significant colour changes were observed in brick, sandstone, weathered sandstone and weathered limestone as the result of pressure washing of either treated or untreated surfaces. This was due to the removal of coloured, weathered surface material or small masonry particles.

As the removal of a sacrificial graffiti barrier requires significant volumes of water to be applied to a surface, this can also cause soluble salt mobilization in weathered masonry where salts are present. The subsequent recrystallization can result in a whitening of the substrate.

The effects of temperature

It was suspected that high pressure hot water or steam might affect surfaces differently from cold water due to differences in solubility and differential expansion effects. Water temperatures of 15–20 °C and 100 °C (59–68 F and 212 °F) were tested. Water temperature changes did not have a significant effect on the samples tested. Hot water was confirmed to be necessary for thorough removal of the sacrificial graffiti barriers.

CONCLUSIONS

Visible colour change

The barriers tested caused detectable colour changes on all surfaces. The level of colour change on fresh and unweathered surfaces was small enough to be considered insignificant. Coatings were more visible on darker surfaces after multiple applications, especially with graffiti residue build-ups. Weathered surfaces also showed a stronger colour change as they allowed more barrier to remain on the surface.

Reversibility

The sacrificial graffiti barriers tested did not penetrate surfaces deeply, but remained within the outer 0–1 mm of the material. While this improved their chances of being substantially removed, none was fully reversible from porous masonry. Residues tended to be left within the recesses of larger pores.

The effect on masonry durability

It was found that all substrate materials showed a decrease in permeability as a result of sacrificial graffiti barrier application. At this stage, therefore, it is logical to assume that these barrier materials will have some pore-blocking effect.

However, even where the application of a sacrificial graffiti barrier causes only a small, subtle reduction in permeability, this can have a very significant effect on salt crystallization within the masonry surfaces. Sacrificial graffiti barriers should not be applied, therefore, to surfaces where there is known to be a soluble salt problem, particularly within a capillary rise zone.

Efficiency of graffiti removal

The barriers were effective in permitting graffiti removal in staged demonstrations by contractors, especially on durable brickwork. However, at water pressures which would not cause damage to the carefully selected surfaces discussed in this report, the degree of graffiti removal was less successful and less consistent. The following points can be made.

- Retreatment of masonry surfaces with sacrificial graffiti barriers reduced the effectiveness of graffiti removal. A reduction in the effectiveness of graffiti removal was found to develop after three applications of the same sacrificial graffiti barrier. Higher water pressures were required to achieve the same degree of removal as progressive applications were made. This means that, as barriers are reapplied, chemical cleaning or the use of higher water pressures would be required for continued effective barrier and graffiti removal. More aggressive cleaning should therefore be anticipated in situations where repeated application of barrier coatings is likely.
- Generally, the sacrificial graffiti barriers allowed better graffiti removal from limestone and terracotta than from the sandstone tested. This was probably due to the nature of the surface pore structure.
- Barriers were not equally effective in terms of graffiti type. One barrier was better at preventing permanent staining with marker pen, while the other was more effective at enabling the removal of spray paint. In addition, the same coatings showed variations in effectiveness with regard to particular graffiti types when used on different substrates. This makes the selection of a universally suitable graffiti barrier impossible.
- Neither sacrificial graffiti barrier tested (SGB1 or SGB3) was able to facilitate complete removal of either spray paint or felt pen graffiti. The level of cleaning achieved was found to be most dependent on the nature of the masonry surface. It was found that the graffiti penetrated through the barriers, staining the masonry underneath, or stained barrier residues which were then not fully removed. Removal of such residual staining would require chemical cleaning or use of unacceptably high water pressures.
- The application of paint on top of a sacrificial graffiti barrier seems to enhance the colour of the paint and hence its visual impact. It is possible therefore that the application of a coating might attract graffiti.
- The application of a sacrificial graffiti barrier does not preclude the use of chemicals for graffiti removal, as is sometimes believed. Chemical cleaning is often required prior to the application of a barrier to remove existing graffiti or to remove graffiti residue.
- Barriers which have been applied over areas cleaned with alkaline materials may be disrupted by soluble salt crystallization of cleaning product residues. Similar breakdowns may also take place if water used has activated an inherent salt loading.
- Some of the greatest damage related to the use of sacrificial graffiti barriers is caused by excessive rinse pressures. Pressure-related damage will take place whether water is hot or cold. Rinse pressures recommended or used by manufacturers are generally higher than those which have been established as safe for use on common historic masonry surfaces (20.7 bar or 300 psi). At these safe, low levels removal of the graffiti cannot be achieved by hot water rinsing alone. The reduction of rinse water pressures is therefore

likely to mean the associated use of chemicals to remove graffiti (and, unavoidably, the sacrificial graffiti barrier beneath) along with the establishment of a frequent re-coating regime.

FUTURE RESEARCH

The work presented here has indicated some of the problems associated with the use of scarificial graffiti barriers on historic masonry, and presented a number of methods for the evaluation of such products on a variety of substrates.

Recommendations for future study include:

- Investigation of the wider range of sacrificial graffiti barriers now available.
- Further assessment of the reversibility of sacrificial graffiti barriers and related substrate changes after multiple treatments.
- Investigation of the use of chemical cleaning methods to remove graffiti and/or barriers from both treated and untreated surfaces. Chemical cleaning of graffiti is not always eliminated by the use of sacrificial graffiti barriers as existing graffiti must be removed before any barriers are applied, and barriers can be ineffective, requiring supplementary cleaning with chemical methods.
- Monitoring of the long-term effects and weathering of sacrificial graffiti barriers.
- Field tests of barrier products and graffiti removal techniques on weathered masonry, which is likely to be the most sensitive to cleaning.

RECOMMENDATIONS

Even in the short term (up to three years), the damage caused by graffiti removal *without* the use of a sacrificial graffiti barrier could be less than or equal to that caused by a similar period of barrier application, removal and reapplication. If graffiti occurs only rarely on a historic building, it will be difficult to justify using sacrificial graffiti barriers for further protection.

If graffiti attacks are undertaken on a regular basis, and chemical cleaning is causing problems, then barriers may be an option. The general use of a sacrificial graffiti barrier on a historic building is unlikely to be appropriate, although it may be the best option for areas of repeated attack in these circumstances.

In the end, prevention is better than cure. If graffiti attacks can be prevented (or at least limited) through the use of management systems such as deterrent landscaping, floodlighting and CCTV, then the potential damage associated with graffiti and graffiti removal will be minimized.

ENDNOTES

1. Sacrificial graffiti barrier 1 = Grafficoat 1 (manufactured by Tensid UK)
Sacrificial graffiti barrier 2: Grafficoat 3 (manufactured by Tensid UK)
Sacrificial graffiti barrier 3: Euregard (manufactured by Delta).
Tensid UK, Aquila House, Wheatash Road, Adlestone, Surrey KT15 2ES; Tel: + 44 1932 564133; Fax: + 44 1932 562546.
Delta AG Ltd, 10 The Butt, Warwick CV34 4SS; Tel: +44 1926 493017; Fax: +44 1926 403711.
2. The pressure water system used was a Kacher HV555 pressure washer.

BIBLIOGRAPHY

British Standards Institute, 1995 *BS 3177: 1959 (1995) Method for Determining the Permeability to Water Vapour Flexible Sheet Materials Used for Packaging*, London.

British Standards Institute, 1997 *BS 6477: 1992 Specification for Water Repellants for Masonry Surfaces*, London.

Lewin, S Z 1982 The mechanism of masonry decay through crystallization, in N Baer (ed), *Conservation of Historic Stone Buildings and Monuments*, 120–144, Washington DC, National Academy Press.

AUTHORS

Nicola Ashurst trained as an architect in Australia and in Rome. Her working career has included six years with the Technical Advisory Service of English Heritage and Adriel Consultancy, a specialist practice providing professional technical input into the cleaning and surface repair of external stone and other traditional masonry materials. In her current work she is actively involved on site in the implementation of conservation, repair and cleaning techniques. She was also involved with the redrafting of BS 6270, the British Standard on cleaning and surface repair of buildings.

Sasha Chapman originally trained as an archaeologist specialising in recording buildings. She joined English Heritage in 1993 and was responsible for coordinating technical enquiries relating to graffiti removal from historic buildings and monuments. She is a former Chair of the United Kingdom Institute for Conservation of Historic and Artistic Works, Stone and Wall Paintings Section.

Susan Macdonald graduated as an architect from the University of Sydney and completed her conservation training at ICCROM, Rome. Susan has worked in private practice in Sydney and the UK. After five years with English Heritage, she is now Principal Heritage Officer: Local Government Heritage Management at the New South Wales Heritage Office in Sydney.

Dr Roy Butlin was manager of the Heritage Support Centre at the Building Research Establishment (BRE) before his retirement. A physical chemist with a background in industry and academe, he worked at the Fire Research Station at BRE before moving into the study of building materials. In 1984 he set up the Weathering Science section at BRE to study the effects of acid deposition on building materials. He retired in 1997.

Matthew Murray studied chemistry at the University of London and Newcastle University. He has been employed by the Building Research Establishment (BRE) since 1992, working on atmospheric pollution damage to buildings, as well as graffiti removal, masonry consolidation and building cleaning using lasers and other methods.

Soft wall capping experiments

HEATHER VILES
School of Geography and the Environment, University of Oxford, Mansfield Road, Oxford OX1 3TB, UK
CHRIS GROVES
Hoffmann Environmental Research Institute, Department of Geography and Geology, Western Kentucky University, 1 Big Red Way, Bowling Green, Kentucky 42101-3576, USA
CHRIS WOOD
English Heritage, 23 Savile Row, London W1S 2ET, UK

ABSTRACT

This chapter examines the results of experiments on the effectiveness of the protective qualities of soft wall capping, with the conservation of historic structures in mind. The main aims of this pilot project were to investigate the thermal blanketing effect of soft wall cappings, and their hydrological impact. Simulations were run under laboratory conditions designed to mimic as far as possible the situation at Hailes Abbey, Gloucestershire.

Key words

Soft wall cappings, soil, vegetation, conservation, water ingress, thermal blanketing

INTRODUCTION

Background to the project

Vegetation growing on ruined masonry walls has long been regarded either as a positive visual feature, the 'romantic ruin', as something to live with or as a problem to be eradicated. English Heritage and its predecessors have usually subscribed to the latter view because of the potential damage caused by roots, greater concentrations of moisture and providing habitats for burrowing animals and other invasive plant species.

English Heritage is responsible for maintaining many miles of ruined walls, most of these being scheduled ancient monuments and therefore among our most precious historic structures. Protecting the walls from the erosive effects of weathering has been a major problem because many of the mortared wall tops of the twentieth century have failed to survive in these very exposed environments, or because their hard impervious nature exacerbated the rate of erosion. Continually repairing wall tops has been an extremely costly exercise and a major drain on restricted resources.

It is claimed that if the hard top works effectively, it channels run-off efficiently, but this concentrates water over parts of the facework, which then suffer undue damage from freeze-thaw cycles. The greatest loss of details seems to follow the most recent repair to the wall-top. The other obvious problem is hard-top cracking (particularly with cementitious mixes), water ingress into the core and the resultant effects of freeze-thaw cycles.

Growing grasses and other plants on wall tops is not new (whether by accident or design). A successful soft top acts as a sponge by absorbing rainfall and then allowing this to evaporate by natural transpiration. Rain falling on the leaves of plants will either sit in droplets if the rain is light (and brief), from where it will evaporate, or, as the droplets build up, they will drip off the leaves into the soil below from where it can be taken up by the roots and transpired by the leaves. Increasingly there is an awareness that soft tops may offer a much more effective, long-lasting and cheap solution to ruined wall protection, as well being more environmentally beneficial. In the last few years soft tops have been applied in a variety of *ad hoc* ways, using different soils and grasses on sites throughout England.

However, no scientific testing or monitoring of sites has been carried out. To date, no work has been done on the physio-chemical effects of the soils and plants on the walls. Besides the obvious question of root damage, nothing has been done to estimate or measure the effects on the weathering mechanisms of rain, frost, diurnal fluctuations in temperatures etc. Although research has been carried out on the chemical weathering of acidic and alkaline rocks, very little of this has been practically applied to the soils and vegetation types selected for soft capping that are currently in place. Little empirical thought has been applied to maintenance regimes or whether damp-proof membranes might be a good thing, or not in such situations.

There is a clear need to research these issues in order to evaluate the efficacy of using soft capping before advice and guidance on the subject can be offered. There is today greater pressure for more 'greening' of monuments, occasioned by renewed interest in the *picturesque*, as well as a desire to protect and enhance the important and unique wall flora for which Britain is particularly renowned in Europe.

As a prelude to designing a full research programme, English Heritage commissioned a short series of laboratory experiments. If successful, these were to form the basis of longer-term experiments which would combine simulated laboratory tests with the monitoring of actual sites and experimental test walls.

The pilot project was jointly designed by English Heritage and the Built Environment Research Group, University of Oxford, and carried out between March

---

* Author for correspondence

Figure 1. Hard wall capping at Hailes Abbey (Photograph by Chris Wood, English Heritage). See Colour Plate 23.

and May 2000. The main aims of the project were to develop and test experimental set-ups to investigate:

- the thermal blanketing effect of soft wall cappings and
- the hydrological impacts of soft wall cappings.

Three sets of experiments were run in order to achieve these aims, which have provided useful data on the performance of three soil and vegetation assemblages. Hailes Abbey, near Cheltenham, Gloucestershire (OS Map 150; SP 050300), was chosen as a basis for the simulations in order to try and relate the experiments to real site conditions.

Hailes Abbey as a case study

Hailes Abbey, run jointly by English Heritage and the National Trust, is a ruined thirteenth-century abbey in the Cotswolds. Built of local Jurassic, oolitic limestone it was destroyed during the Dissolution in the sixteenth century, when many ecclesiastical buildings were destroyed or ruined. In recent years, frost weathering has been a major problem, as the site is situated in a frost hollow, and both hard and soft wall capping have been carried out (Figs 1 and 2). Details on the soft wall capping techniques used have been hard to come by, but an examination of the site showed the capping to consist of turf placed on thin soils with an underlying permeable membrane.

Figure 2. Soft wall capping at Hailes Abbey (Photograph by Heather Viles). See Colour Plate 24.

METHODOLOGY

Project design

In order to fulfil the aims of the project, to develop and test experimental set-ups to investigate the thermal blanketing effect of soft wall cappings and the hydrological impacts of soft wall cappings, a sequence of steps was followed:

- selection of materials
- selection of climatic data for simulations
- design and testing of experimental set-ups
- running experiments
- analysis of results

The thermal blanketing experiments were designed and run first, followed by the hydrological experiments, which were divided into two sections, ie water penetration and soil water-holding experiments. Regular progress meetings were held for all members of the research team in order to critically assess each step of the project.

Selection of materials and climatic data for all experiments

In order to simulate the Hailes Abbey environment the following materials and climatic information were used:

- Stone from Brockhill Quarry at Naunton, of the same type as has been used at Hailes Abbey for recent hard wall capping (from Grange Hill Quarry), was used to represent the stone under the experimental soft wall capping and hard wall caps. The stone was cut into 240 x 240 x 30 mm (9.5 x 9.5 x 1 $\frac{1}{8}$ in) blocks for the experiments. A detailed description of the stone is provided in Annex A.
- Mixed species of grass, including fescues and rye grass, from an established, well-managed garden on limestone from the Woodstock area, were used as turf in all experiments. Further information on the grass is given in Annex B.
- Three soil types were used, one from a garden on the Oxford river terraces (soil A), one from the garden which provided the turf (soil B) and one made

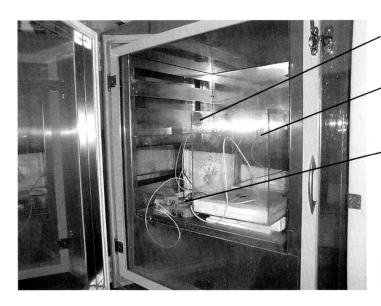

Figure 3. The Fisons Environmental Cabinet, showing the inside with thermal blanketing experiment running (Photograph by Heather Viles).

specially for the project using a mixture of sand and peat (soil C). The soils are described in further detail in Annex B.
- Two soil thicknesses were used (100 and 200 mm: 4 in and 8 in), based on informal discussions with English Heritage and observations at Hailes Abbey.
- Climatic data from the Oxford Radcliffe Meteorological site (run by the School of Geography and the Environment, University of Oxford) was consulted to provide meaningful temperature and rainfall inputs to the experiments, and also to compare with the soil and stone surface temperatures produced in the experiments. (The Radcliffe Meteorological Station collected daily minimum concrete surface, grass and soil temperatures, see Annex C.)

THERMAL BLANKETING EXPERIMENTS

Experimental design

A series of experiments was designed to test the thermal blanketing role of soft wall cappings of different depths and soil types, in comparison with hard wall caps of different thicknesses and in comparison with bare stone. Comparisons were also made between the performance

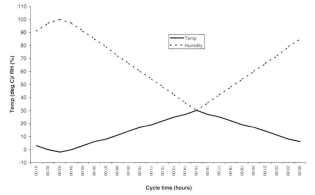

Figure 4. Graph of the temperature and humidity cycle programme used in the Fisons Environmental Cabinet.

of wet and dry soft and hard caps. The general procedure involved measurement of temperatures on the stone surfaces and within the soils under carefully controlled conditions within a Fisons environmental cabinet. There were three basic components of the experiment.

The environmental cabinet was designed to produce controlled cycling of temperature and humidity and is programmable to cover a wide range of possible climatic situations (Fig 3). The cabinet was programmed using climatic data from Oxford to give a synthetic 24-hour cycle which went from a cold January night to a hot July

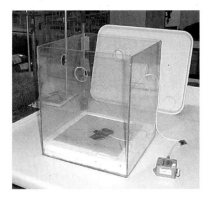

Figure 5. Design of experimental boxes for thermal blanketing experiments: bare stone with rock surface temperature probe (Photograph by Heather Viles).

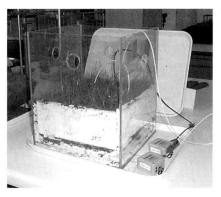

Figure 6. 100 mm (4 in) soil over stone with rock surface probe on stone surface and one pencil tip probe at 50 mm (2 in) depth in soil (Photograph by Heather Viles).

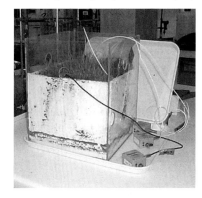

Figure 7. 200 mm (8 in) soil over stone with rock surface probe on stone surface and two pencil tip probes (one at 50 mm and one at 100 mm depth in soil: 2 in and 4 in) (Photograph by Heather Viles).

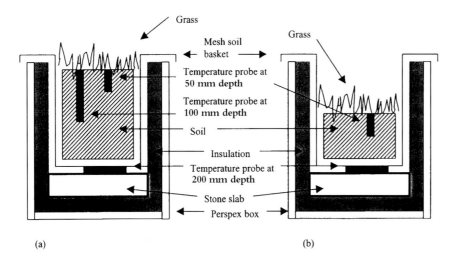

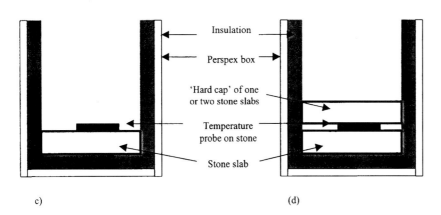

Figure 8. Temperature blanketing experiments design and monitoring: (a) 200 mm thick soil (8 in); (b) 100 mm thick soil (4 in); (c) bare stone; (d) hard cap.

day (using Radcliffe Meteorological Station data from 1999). The cycle is shown in Figure 4.

The experimental boxes were designed and built specially for this project (Figs 5, 6 and 7). They consist of 300 mm³ (1 ft³) perspex boxes (to provide a rigid outer casing) with holes drilled to encourage air flow. The boxes have an inner insulating layer of polystyrene (20–30 mm, ³⁄₄–1¹⁄₈ thick) to prevent loss of heat through the sides and base of the soil and stone. The blocks of Brockhill stone were placed at the base of each box, on top of the insulating polystyrene. Soft caps were created by cutting a square of turf, placing it on a block of soil (except in the case of soil B where the turf was cut on site to incorporate 100 and 200 mm (4 in and 8 in) depths of soil beneath) and placing both in a plastic mesh bag (5 x 5 mm or ¹⁄₄ ins mesh), to prevent loss of material during making up the test boxes and to create a permeable membrane at the base of the soil. Hard caps were created by placing one (30 mm cap: c 1 ¹⁄₈ in) or two (60 mm cap: 2 ³⁄₈ in) blocks of Brockhill stone on the base stone.

The temperature monitoring during the experimental runs was carried out using a suite of Tinytag data loggers connected to temperature and relative humidity probes. These probes and loggers are robust, small and reliable and can be programmed to collect data over a range of timespans at varying intervals. For these experiments the loggers recorded data at five minute intervals. The monitoring scheme is shown in Figure 8 and involved using Gemini surface temperature probes to monitor temperature on the surface of the stone blocks and Gemini sheathed thermistor probes to monitor temperatures within the soil at varying depths: usually at 50 mm (2 in) in the 100 mm (4 in) soil and at 50 mm (2 in) and 100 mm (4 in) in the 200 mm (8 in) soil. A Gemini relative humidity probe monitored the humidity within the environmental cabinet, and a Digitron air temperature sensor connected to another datalogger monitored the air temperatures within the cabinet.

Experimental runs

Tables 1 and 2 summarize the entire sequence of experimental runs carried out, including the initial ones to test the equipment and develop good protocols which would produce meaningful results. Preliminary experimental runs showed that in order to get a meaningful 24-hour cycle each experiment needed to be run for 30–31 hours to allow the boxes to settle down to the temperature inside the cabinet. In each of the main experiments listed in Table 2, three boxes were placed in the environmental cabinet at one time.

Major results

Because of the uniformity of the conditions in the environmental cabinet between runs, and given a 6–7 hour equilibration time at the beginning of each experi-

Table 1. *Experiment Set A: Exploratory runs and soil A thermal blanketing effects.*

Run	Begin date	Cycle time[†]	Duration	Treatments	Observed Parameters
A	14/3/00	0800-2400	16 hours	Test probe operation	surface and soil temperature; wetness; humidity
B	15/3/00	0800-2400	16 hours	Box 1: 100 mm soil A and grass★	soil surface & 100 mm temperature (middle and edge); humidity
C	16/3/00	0800-2400	16 hours	Box 1: bare stone	soil surface & 100 mm temperature (middle and edge); humidity
				Box 2: 100 mm soil A over 30 mm stone	stone surface temperature (middle and edge)
D	17/3/00	0300-0748	5 hours	Box 1: bare stone	stone surface temperature; humidity
				Environmental cabinet	humidity
E	17/3/00	0800-2400	64 hours	Box 1: bare stone	soil surface temperature; 100 mm soil temperature (middle and edge); humidity (first 24 hrs)
				Box 2: 100 mm soil A and over 30 mm[†] stone	stone surface temperature; humidity (next 40 hrs)
				Environmental cabinet	temperature
F	21/3/00	0700-2300 (BST begins)	16 hours	Box 1: bare stone (dry)	stone surface temperature
				Box 2: 100 mm soil A over stone (dry)	soil surface temperature; 100 mm soil temperature
				Box 3: 200 mm soil A over stone (dry)	soil surface temperature; 100 mm soil temperature
				Environmental cabinet	humidity
G	22/3/00	0700-2300	16 hours	Box 1: bare stone (wet)	stone surface temperature
				Box 2: 100 mm soil A over stone (wet)	soil surface temperature; 100 mm soil temperature
				Box 3: 200 mm soil A over stone (wet)	soil surface temperature; 100 mm soil temperature
				Environmental cabinet	humidity
H	23/3/00	0700-2300	16 hours	Box 1: bare stone (dry)	stone surface temperature
				Box 2: 100 mm soil A over stone (dry)	50 mm and 100 mm soil temperature
				Box 3: 200 mm soil A over stone (dry)	50 mm, 100 mm, and 200 mm soil temperature
				Environmental cabinet	humidity
I	24/3/00	0700-2300	64 hours	Box 1: bare stone (wet)	stone surface temperature
				Box 2: 100 mm soil A over stone (wet)	50 mm and 100 mm soil temperature
				Box 3: 200 mm soil A over stone (wet)	50 mm, 100 mm, and 200 mm soil temperature
				Environmental cabinet	humidity

[†] time with respect to 24 hour cycle within environmental chamber
★all soils used in experiments include live grass cover

Table 2. *Experiment Set B: Hard cap and soft cap thermal blanketing effects.*

Run	Begin date	Cycle time[†]	Duration	Treatments	Observed Parameters
J	27/3/00	0700-1300	30 hours	Box 1: bare stone (dry)	stone surface temperature
				Box 2: 30 mm stone hard cap over stone (dry)	stone temperature beneath 30 mm stone hard cap
				Box 3: 60 mm stone hard cap over stone (dry)	stone temperature beneath 60 mm stone hard cap; polystyrene foam insulation temperature
				Environmental cabinet	humidity
K	29/3/00	1500-2200	31 hours	Box 1: bare stone (wet)	stone surface temperature
				Box 2: 30 mm stone hard cap over stone (wet)	stone temperature beneath 30 mm stone hard cap
				Box 3: 60 mm stone hard cap over stone (wet)	stone temperature beneath 60 mm stone hard cap; polystyrene foam insulation temperature
				Environmental cabinet	humidity
L	30/3/00	2400-0700	31 hours	Box 1: 30 mm stone hard cap over stone (dry)	stone temperature beneath 30 mm stone hard cap
				Box 2: 100 mm soil B over stone (dry)	50 mm and 100 mm soil temperature
				Box 3: 200 mm soil B over stone (dry)	50 mm, 100 mm, and 200 mm soil temperature
				Environmental cabinet	humidity
M	31/3/00	0900-1600	31 hours	Box 1: 30 mm stone hard cap over stone (wet)	stone temperature beneath 30 mm stone hard cap
				Box 2: 100 mm soil B over stone (wet)	50 mm and 100 mm soil temperature
				Box 3: 200 mm soil B over stone (wet)	50 mm, 100 mm, and 200 mm soil temperature
				Environmental cabinet	humidity
N	2/4/00	2300-0600	31 hours	Box 1: 30 mm stone hard cap over stone (dry)	stone temperature beneath 30 mm stone hard cap
				Box 2: 100 mm soil C over stone (dry)	50 mm and 100 mm soil temperature
				Box 3: 200 mm soil C over stone (dry)	50 mm, 100 mm, and 200 mm soil temperature
				Environmental cabinet	humidity
O	3/4/00	0800-1500	31 hours	Box 1: 30 mm stone hard cap over stone (wet)	stone temperature beneath 30 mm stone hard cap
				Box 2: 100 mm soil C over stone (wet)	50 mm and 100 mm soil temperature
				Box 3: 200 mm soil C over stone (wet)	50 mm, 100 mm, and 200 mm soil temperature
				Environmental cabinet	humidity

[†] time with respect to 24 hour cycle within environmental chamber
★all soils used in experiments include live grass cover

Table 3. Data sets from thermal blanketing experiments

Set-up	Dry	Wet
Bare stone	Run J, Box 1	Run N, box 1
	Run K, box 1	Run O, box 1
30 mm hard cap	Run J, box 2	Run K, box 2
	Run L, box 1	Run M, box 1
60 mm hard cap	Run J, box 3	Run K, box 3
100 mm soft cap, soil B	Run L, Box 2	Run M, box 2
200 mm soft cap, soil B	Run L, box 3	Run M, box 3
100 mm soft cap, soil C	Run N, box 2	Run O, box 2
200 mm soft cap, soil C	Run N, box 3	Run O, box 3

ment it is possible to compare the results of each box in the six experiments. Thus, we have 18 different sets of data to compare, as described in Table 3.

This matrix of data permits comparison between soft caps of different thickness, soft caps of differing soil types, soft caps of differing wetness and between soft and hard caps, and both cap types and bare stone. Figures 9 to 14 show the results of the major experimental runs and Table 4 presents a summary. Both soft caps and hard caps produced a thermal blanketing effect.

- The soil caps produced a lowering of maximum stone surface temperatures of between 4.5 and 7 °C (40 and 44 °F), accompanied by a lag in the timing of the maximum temperature in comparison with the bare stone of up to three hours.
- **Thickness of soil cap (B and C)**: In all cases (wet and dry, both soil types) the thicker soil cap produced a lower temperature maximum (of 1–2 °C (33.8 – 35.6 °F). In all but one case the thicker cap also produced a higher minimum temperature (by 0.5–3.4 °C (32.9–38 °F)). The lag in maximum and minimum temperatures experienced by the subsurface stones in comparison with the bare stone was

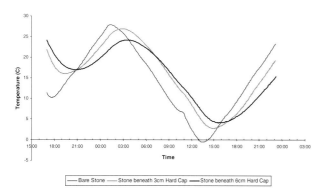

Figure 9. Results from Run J: dry hard cap.

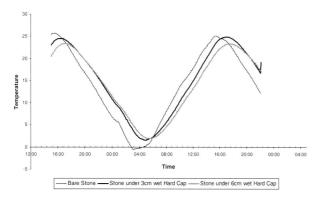

Figure 10. Results from Run K: wet hard cap.

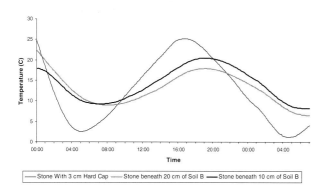

Figure 11. Results from Run L: dry hard cap and dry soil B cover.

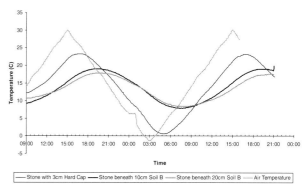

Figure 12. Results from Run M: wet hard cap and wet soil B cover.

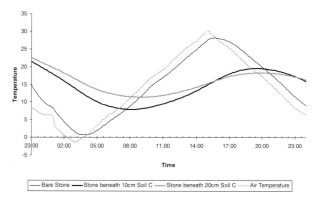

Figure 13. Results from Run N: dry soil cover.

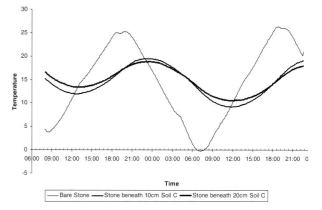

Figure 14. Results from Run O: wet soil cover.

Table 4. Data summary from thermal blanketing experiments

Run/box	Max temp °C	Time of max. temp	Min temp°C	Time of min. temp
Run J				
Bare stone dry	28		-1	
30 mm hard cap dry	27		3	
60 mm hard cap dry	24		4	
Run K				
Bare stone wet	25		-1	
30 mm hard cap wet	25		2	
60 mm hard cap wet	23		2.5	
Run L				
30 mm hard cap dry	25.2	2:35	1.1	14:30
100 mm soil B dry	20.4	4:25	8.2	16:10
200 mm soil B dry	18	5:00	6.4	17:00
Run M				
30 mm hard cap wet	23.5	2:45	0.6	14:55
100 mm soil B wet	19	4:50	8	17:20
200 mm soil B wet	18	5:15	8.5	17:05
Run N				
Bare stone dry	28.2	15:40	0.73	3:40
100 mm soil C dry	19.6	19:20	8	8:05
200 mm soil C dry	18.3	19:30	11.4	8:55
Run O				
Bare stone wet	25.2	19:55	-0.2	6:55
100 mm soil C wet	19.5	23:10	9.2	11:40
200 mm soil C wet	18.8	23:00	10.5	11:50

NB: These times are cycle times. Those on the X-axes in Figs 9–14 are real times.

approximately the same, with variation between runs as to whether the thick or thin cap responded quickest.
- **Wetness of soil cap (B and C)**: Although the wet bare stone and wet hard cap showed a clear lowering in maximum temperature (and a similar lowering of minimum temperatures) the differences between wet and dry soil caps were less clear-cut.
- **Soil B vs soil C**: Soils B and C showed a similar blanketing effect, although soil C created higher minimum temperatures than soil B and also imposed a much longer lag time on both heating and cooling in comparison to the bare stones.
- **Soils vs hard caps**: Both soil types and thicknesses provided a greater thermal blanketing effect than the 30 mm and 60 mm (1 $^1/_8$ in and 2 $^3/_8$ in) hard caps.

Evaluation of the experimental design

As these are pioneering experiments there were no standard protocols to follow and the experimental design had to be developed to fit the project. Some suggestions for future research are:

- The programmed cycle (from January night to July day) is fairly extreme, but the lowest point of the cycle does not give as low surface and sub-surface temperatures as recorded from Oxford in 1999 (see Appendix 3). The Oxford data showed that grass and concrete surface temperatures both plunged below −5 °C (23 °F) on three occasions, although at depths of 50 mm (2 in) in the soil temperatures only went below 0 °C (23 °F) once during the year.
- The environmental cabinet was not found to be able to reproduce the desired relative humidity cycle alongside the temperature cycle. Improvements to the cycle need to be made if relative humidity levels are to reflect reality.
- The hard cap simulation was not as realistic as desired and could be improved upon, given time to build a proper hard cap using mortar and allowing it to settle.

WATER PENETRATION EXPERIMENTS

Experimental design

A series of experiments was designed to test how long it took for rainfall to penetrate through the soft cap and reach the underlying stone, as well as to investigate the nature of water flow through the soil (ie whether it was concentrated into particular pathways). Intuitively, one would think that coarser textured soils or those containing more macropores and pipes (where macropores are those over 60 microns in diameter, and pipes are well-connected pores systems of routeways such as root channels) would show more quick throughflow of water. Water being channelled down onto the underlying stonework in this fashion would have the potential to cause physical and chemical damage. If the soil was reasonably acidic then water flowing through the soil might become more acidic and thus be capable of damaging limestone, for example. Other soils might be conceived of as being more like a sponge, soaking up water from a storm event and preventing it reaching the underlying stonework.

In order to test these ideas a modified version of the perspex soil box was rained on using a rainfall simulator as shown in Figure 15. The basic components of these experiments were:

- The rainfall simulator was a portable simulator built by John Morgan for soil hydrology experiments carried out some years ago in the School of Geogra-

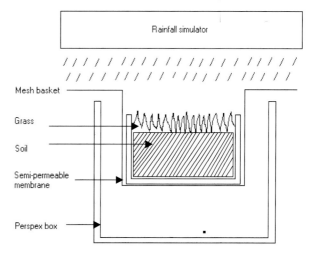

Figure 15. Experimental set-up used for water penetration experiments. The aim was to investigate whether soft caps protected the underlying stone from rain water. The rainfall simulator was run for up to three hours. Any throughflow water was observed using graph paper at the base of the perspex box, or captured for chemical analysis. There was no stone in these experiments. Water chemistry of throughflow water was measured to assess its 'potential' for dissolving limestone.

Figure 16. The rainfall simulator, showing the reservoir (top right), the water pressure control system (directly below reservoir) and the nozzles and droplet formers (at top of cage) (Photograph by Heather Viles).

Figure 17. The rainfall simulator, showing the nozzles and droplet formers (at top of cage) (Photograph by Heather Viles).

Figure 18. The rainfall simulator, showing the basket with soil and permeable cloth wrap (Photograph by Heather Viles).

phy (Figs 16, 17 and 18). A water tank on the top provides a reservoir and head, and then a simple manometer device is used to manipulate and control flow rates through a capillary tube. The water then fills a perforated container underneath and flows out through a series of small holes. A wire mesh underneath provides an additional baffle, and helps to reduce raindrop size. The simulator has the advantage that it is small, easy to assemble and required no construction costs. In theory it can use deionized water or rainwater (either natural or artificial). In practice tap water has to be used to overcome some problems with regulating flow rates. The disadvantages proved to be that it required a lot of maintenance and fine-tuning in order to ensure realistic rainfall intensities. Using Radcliffe Meteorological Station data from 1998 on the duration and total rainfall of characteristic rain storms, some realistic rainfall intensities were derived (see Table 5).

- The experimental boxes: the same 300 mm^3 (1 ft^3) perspex boxes were used, but this time soil with overlying grass turf was placed in a wire mesh basket (sometimes contained within a permeable cloth suspended within the perspex box without stone underneath). This allowed us to observe the outflow of water from the bottom of the soil and to collect the water running through the soil for subsequent chemical analysis.
- The observation of throughflow was carried out by eye (to observe the timing and nature of water flow out from the base of the soil), by destructive sampling in the form of cutting open the soil profile (to observe the depth and nature of wetting through soils when no water came out of the base) and by chemical analysis (to investigate the evolution of the soil water in terms of acidity and major ion contents).

Experimental runs

Table 6 gives a summary of the experimental runs carried out, some of which were aimed at testing the set-up and some which were aimed at providing initial data. Only in runs G, K and L was it possible to produce rainfall amounts which were realistic (although even in these runs the intensity was an order of magnitude higher than found in heavy rain in Oxford).

Table 5. Summary of Oxford rain storm characteristics, 1998.

Month (1998)	Number of storms	Mean precipitation (mm)	Mean duration (hrs)	Mean intensity (mm/hr)	Mean time since preceding rain (hrs)
January	35	1.99	3.29	0.42	16.66
February	6	2.57	4.57	0.87	195.14
March	34	1.918	3.294	0.477	17.441
April	61	1.97	2.26	0.70	9.67
May	17	1.41	1.76	0.83	43.47
June	47	1.81	2.13	0.61	12.83
July	24	1.33	1.75	0.52	30.08
August	16	1.81	2.5	0.48	44.31
September	36	3.98	2.89	1.07	11.17
October	45	2.43	2.67	0.62	11.84
November	31	2.43	2.94	0.59	16.16
December	40	3.90	5.07	0.98	32.59

Table 6. Runs of water penetration experiments.

Run	Treatment	Intensity (mm/hr) or volume (mL)	Purpose
A	100 mm dry soil B, dry grass	108 mm/hr	Test experimental setup
B	100 mm dry soil B, dry grass	108 mm/hr	Test experimental setup, soil B hydrology under intense rain
C	Rainwater collected straight from drips	60 mm/hr	Background tapwater chemistry
D	Perspex box with empty wire cage	60 mm/hr	Chemistry and hydrology effect from experimental setup
E	Perspex box with plastic coated wire cage	60 mm/hr	Chemistry and hydrology effect from experimental setup
F	Perspex box with wire basket and black cloth	60 mm/hr	Chemistry and hydrology effect from experimental setup
G	100 mm dry soil C, dry grass	15 mm/hr	Dry soil C and grass hydrology under realistic rainfall conditions
H	Bare limestone	200 mL	Limestone hydrology & chemical effect on tap water (5 min)
I	Bare limestone	200 mL	Limestone hydrology & chemical effect on tap water (20 min)
J	Bare limestone	200 mL	Limestone hydrology & chemical effect on tap water (60 min)
K	100 mm dry soil B, dry grass	18 mm/hr	Dry soil B and grass hydrology under realistic rainfall conditions
L	100 mm dry soil C, no grass	18 mm/hr	Dry soil B (no grass) hydrology under realistic rainfall conditions

Major results

Run B

Using high rainfall intensities of 108 mm/hr (4 1/4 in/hr) onto 100 mm (4 in) thick, dry soil B without the permeable cloth around the soil, water came through the bottom of the soil after three minutes and 20 seconds, in the form of a group of three steadily dripping spots. When the rainfall simulator was finally turned off, it took a long time for water to stop percolating out of the bottom of the soil.

Run G

Using lower rainfall intensities of 15 mm/hr (5/8 in/hr) onto 100 mm (4 in) thick dry soil C with the permeable cloth around the soil, three hours of rainfall produced no water flow out of the bottom of the soil. Post-experimental destructive sampling (which involved cutting the soil in half vertically towards the middle to view the wetting front) showed that the soil was damp in the centre all the way to the bottom, but with no evidence of saturation at any point down the profile (Fig 19). The largest rain storm recorded in Oxford in 1998 was 49 mm

Figure 19. Soil cross-section after run G of the water penetration experiments (Photograph by Chris Groves).

Figure 20. Soil cross-section after run K of the water penetration experiments (Photograph by Chris Groves).

(2 in) falling in 20 hours. The simulation applied 54 mm (2 1/8 in) in three hours. This result suggests that most rainstorms in Oxford would not be capable of producing runoff at the base of the 100 mm (4 in) soil specimen.

Run K
Using rainfall intensities of 18 mm/hr (c 3/4 in/hr) on dry soil B with the permeable cloth we found that three hours of rain again produced no measurable flow from the base of the soil, but showed that some saturated areas were present (Fig 20). Clear vertical pathways for waterflow could be seen in the soil.

Run L
Using rainfall intensities of 18 mm/hr (c 3/4 in/hr) on dry soil C, this time with no grass cover but with the permeable cloth, three hours of rain again produced no visible water flow from the base of the soil, but as Figure 21 shows, the basal layer was almost saturated.

Evaluation of experimental design

There were several problems with this series of experiments which related mainly to the rainfall simulator. For future experiments it is necessary to:

- create more realistic rainfall intensities using a better rainfall simulator. During one thunderstorm, on 17/4/00, rain and hail were collected in a rain gauge and found that 0.8 mm fell in four minutes, giving an intensity of 12 mm/hr (c 1/2 in/hr). Thus, short-term rain events may be as intense as our simulated intensities, although the mean intensity over a whole storm event (Table 5) is often much lower.
- use natural rainwater where possible if chemical analyses of the water percolating through the soil are to be attempted
- use a range of different permeable membranes around the soil to test their impact on water flow rates and routes.

SOIL WATER-HOLDING EXPERIMENT

Experimental design

An experiment was designed to investigate the water-holding capacity of soft wall cappings. Once saturated it could be suggested that soils will ensure that the stone beneath them is kept moist (and therefore vulnerable to both physical and chemical attack) for much longer than bare stone. Three experimental set-ups were built, one

Figure 21. Soil cross-section after run L of the water penetration experiments (Photograph by Chris Groves).

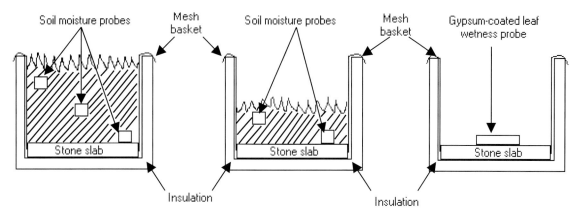

Figure 22. Drying of soils and bare stone surface during soil water-holding experiment. The aim was to investigate the water-holding qualities of soft caps. Three test baskets were left on a roof, open to the elements and the soil was saturated. They were left to dry out naturally over the course of a few weeks. Daily measurements of soil moisture were taken. The leaf wetness probe recorded every five minutes and was downloaded weekly.

with bare stone, one with 100 mm (4 in) deep soil C and grass and one with 200 mm (8 in) deep soil C and grass cover. The components of the experimental set-up were as follows:

- In this case it was felt that perspex boxes would prevent natural drying conditions, and so a box was constructed from polystyrene insulation material around the metal mesh basket used in the water penetration experiments above. The polystyrene was firmly taped together at the edges, and perforated with many holes to allow some airflow through the soils, while still providing some shelter and insulation.
- The wetting and drying was carried out by using a spray to apply enough water to saturate the soil. As interest was in drying rather than wetting, in this experiment it was not felt necessary to replicate natural rainfall conditions. However, it was important to apply the same amount of water to each experimental set-up. Water was applied equivalent to a 20 mm ($c^3/_4$ in) storm over a 15 minute period. In order to simulate natural drying conditions as much as possible, the experimental boxes were left on the roof of the School of Geography and the Environment in a semi-sheltered position, open to both sun and rain.
- Monitoring of soil moisture levels was done using an ELE portable soil moisture meter with glass fibre and metal soil moisture cells (Fig 22). The moisture cells were placed within the soil profiles at 50 and 100 mm (2 and 4 in) in the 100 mm (4 in) soil and at 50, 100 and 200 mm (2, 4 and 8 in) in the 200 mm (8 in) soil and readings taken at intervals of a few hours to several days. Monitoring of bare stone surface moisture was done using a modified leaf wetness probe (prepared according to the methods of See et al 1988) connected to a Tinytag datalogger which collected data every five minutes. Both moisture measurement methods require calibration after the experiment. Data on air temperature, wind and rain over the period of the experiment was obtained from the Radcliffe Meteorological Station

The boxes were placed on the roof and left for 35 days.

Preliminary results

Figure 23 gives the main results of the soil water-holding experiment. It can be seen that both soils, once saturated, took an extremely long time to dry out. The 100 mm (4 in) soil began to dry out noticeably after 15 days. A few days later a period of heavy rain affected Oxford, re-wetting the soils. The 200 mm (8 in) soil showed no evidence of drying out within the timescale of the experiment. In contrast, the bare stone surface showed repeated cycles of wetting and drying, probably as a result of condensation of small amounts of water on the modified leaf wetness probe after each night.

Evaluation of the experimental design

In an experiment of this sort it is difficult to obtain representative conditions using small blocks of soil and grass, and finding a suitable controlled environment (without additional rain inputs) is also a challenge. The two types of probes used (soil moisture cells and modified leaf wetness probe) are difficult to calibrate and it is hard to compare the results from the two techniques. These are surmountable problems, however, and future studies should allow some preliminary time to improve upon these measurement techniques.

DISCUSSION

The suite of experiments documented in this report have produced some useful pilot results. In each case the experimental design has proved to be robust and effective, although, in all cases, some improvements in the realism and repeatability of the experiments could be made. Some general suggestions for further work are:

- Longer runs of each type of experiment would provide better results.
- Better data collected from monitoring soft wall capping in the field could be used in a coordinated fashion

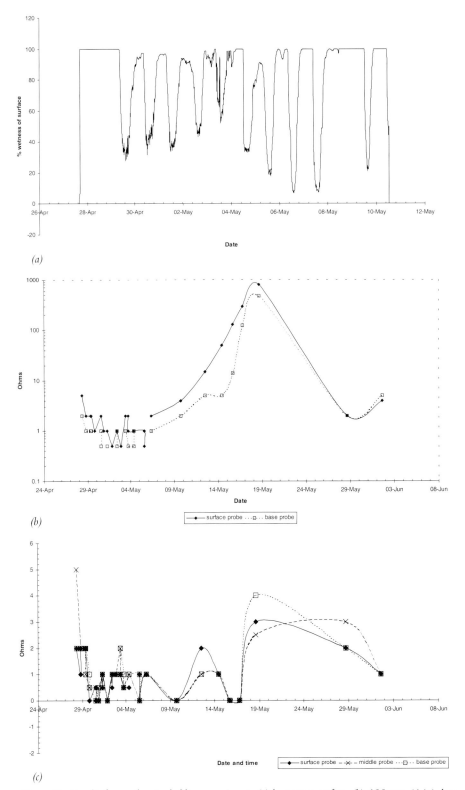

Figure 23. Results from soil water-holding experiment: (a) bare stone surface, (b) 100 mm (4 in) deep soil, (c) 200 mm (8 in) deep soil.

with laboratory experimentation to provide more realistic simulations.

- A longer, integrated programme of field and laboratory studies is desirable to provide a good, balanced assessment of the roles of soft wall cappings of different soil types, involving different vegetation assemblages and overlying different types and geologies of walls.

ANNEX A: CHARACTERIZING THE STONE

Methods

Two samples of freshly quarried and cut stone from Brockhill Quarry, Naunton (OS Map 45; SP 133238), were analysed using an SEM (Scanning Electron Microscope) as well as two samples of frost-damaged stone

Large hollow caused by detachment of c 750 μm ooid

Figure 24. Frost-weathered stone from Hailes Abbey at ground level (Photograph by Heather Viles).

Ooid c 250 μm in diameter

Figure 25. Freshly quarried stone from Brockhill Quarry, Naunton (Photograph by Heather Viles).

removed from Hailes Abbey (at ground level). Each sample was fractured from a larger block, glued to an aluminium stub, gold-coated and observed with a Cambridge Stereoscan 90 SEM.

Results

Figures 24 and 25 give representative, low magnification views (x 76) of both stones. The frost-damaged stone from Hailes Abbey is an oolitic limestone with ooids of between 500 μm and 1 mm in diameter surrounded by a matrix of cement. The freshly quarried stone is also oolitic in structure, the only clear difference being that there are fewer ooids, and they are smaller (generally c 250 μm in diameter).

ANNEX B: CHARACTERIZING THE SOIL AND VEGETATION

Methods

Detailed soil analyses have been carried out as follows:

- colour (using the Munsell Soil Colour Chart)
- grain size distribution (by sieving and granulometry)
- pH (with standard laboratory pH meter)
- bulk density.

Table 7. Colour analyses.

Soil	Munsell Notation	Description
Soil A	10YR 2/1	Black
Soil B	10YR 3/3	Dark brown
Soil C	10 YR 3/2	Very dark greyish brown

Table 8. Grain size distribution analyses. % fines = % of sample by weight of <63 microns in diameter. Mean and standard deviation are expressed in Phi units and are derived graphically from the > 63 micron fraction. The standard deviation gives a measure of the degree of sorting of sediment.

Soil	% fines	Mean grain size	Standard deviation
Soil A	8.88	1.02	1.60
Soil B	9.56	-0.53	1.11
Soil C	7.10	0.6	1.36

Table 9. pH analyses.

Soil	PH
Soil A	6.3
Soil B	7.3
Soil C	6.2

Table 10. Bulk density analyses, measurements expressed in kg/m^3.

Soil	Dry bulk density	Wet bulk density
A	444	592
B	312	425
C	284	353

Table 11. Summary statistics of surface temperature data (Radcliffe Meteorological Station, Oxford, 1999).

Site	Mean °C	Standard deviation	Maximum °C	Minimum °C	Range
Grass	4.7	4.9	16.4	-6.2	22.6
Concrete	6.2	5.2	17.9	-6	23.9
50 mm soil	11.5	6.6	26.7	-0.4	27.1
100 mm soil	10.9	5.8	22.5	0.5	22
200 mm soil	11.6	5.6	23	1.9	21.1

Simple visual observations have been made of the vegetation cover (above and below the soil).

Results

Colour (Table 7)
GRAIN SIZE DISTRIBUTION (TABLE 8)
Soil A has the smallest average grain size (c 0.5 mm) and soil B the largest (c 1.41 mm) with soil C at 0.71 mm. These all fall within the coarse category. From the standard deviations we can tell that soil B is moderately sorted, with soils A and C poorly sorted.

pH (Table 9)
Both soils A and C are thus slightly acid, with soil B neutral.

Bulk density (Table 10)
Bulk density is the mass of sediment per unit volume. It relates closely to infiltration capacity and is controlled by particle density, organic matter contents, porosity and moisture contents. Soil A has the highest bulk density, and thus the best structure, followed by soil B with soil C having the lowest values.

Vegetation
The vegetation used for all samples came from the top of soil B and was a low diversity short turf. Immediately under the surface of the soil was a layer of thick, dense root mat (c 20 mm thick: $c\ ^3/_4$ in] underlain by a mid-density root zone which extended to the base of the 100 mm (4 in) thick slabs. All roots were very fine (< 1 mm in diameter). At the end of the long-term drying experiment when the blocks were exhumed it was found that a dense root mat had developed all the way to the base of both 100 and 200 mm (4 and 8 in) blocks of soil C.

ANNEX C: CLIMATIC DATA

Climatic data were obtained from the Radcliffe Meteorological Station, Oxford for 1999. Hourly air temperature and relative humidity data were analysed for January

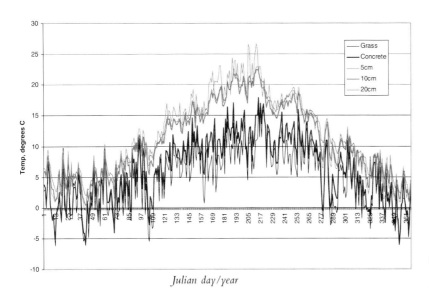

Figure 26. Daily minimum temperatures on grass and concrete surfaces and at 50, 100 and 200 mm (2, 4 and 8 in) soil depths (Radcliffe Meteorological Station, Oxford, 1999).

and July 1999 to provide a synthetic dataset representing a cycle from cold winter conditions at night to warm summer conditions during the day. These data were used to programme the environmental cabinet. Because of a slight gap in the programme, a small 'thermal plateau' actually occurred in the cycling.

Daily minimum temperatures were also obtained for grass and concrete surfaces as well as at 50, 100 and 200 mm (2, 4 and 8 in) depths in soil, as presented in Figure 26 and summarized in Table 11. These data suggest that our synthetic cycle provides a reasonably harsh test of minimum temperatures. The data will also be used later in the study to compare with the experimental results.

EQUIPMENT

environmental cabinet: model FE/300/MP/R20, from Sanyo Gallencamp plc, Monarch Way, Belton Park, Loughborough, LE11 5XG, UK.

scanning electron microscope: Stereoscan 90, from Cambridge Instruments, Viking Way, Bar Hill, Cambridge CB3 8EL, UK.

pH meter: Philips PW9418/10, from Philips House, Cambridge Business Park, Cowley Road, Cambridge CB4 0HA, UK.

ELE moisture meter (soil): MC-305B, from ELE International Ltd, Eastman Way, Hemel Hempstead HP2 7AB, UK.

Munsell Soil Colour Chart, from GretagMacbeth, 617 Little Britain Road, New Windsor, NY 12553, USA.

Gemini pencil probe, Tiny Tag Plus temperature high resolution, Tiny Tag Plus relative humidity, Tiny Tag wetness and Tiny Tag Plus, from Gemini Data Loggers UK Ltd, Scientific House, Terminus Road, Chichester PO19 2UJ, UK.

Digitron air temperature probe: TK SFL, from Sifam Instruments Ltd, Woodland Road, Torquay TQ2 7AY, UK.

BIBLIOGRAPHY

See R B, Reddy M M and Martin RG, 1988 Description and testing of three moisture sensors for measuring surface wetness on carbonate building stones, in *Review of Scientific Instruments* **59**:10, 2279–84.

ACKNOWLEDGMENTS

The authors would like to acknowledge the help and support provided by Niall Morrissey of English Heritage, Professor Andrew Goudie, Chris Jackson, Pat and Albert Woodward, Sue Chitty, Nick Carter, Clare Cox, Dave Banfield, John Morgan, the workshop of the Department of Engineering, University of Oxford, and Julian Palmer, Brockhill Quarry. The experiments were largely run by Nia Dadson, Julia Holgate, Johnny Merideth and Mike Perring, and Sarah Antill helped design and construct the experimental set-up.

AUTHOR BIOGRAPHIES

Heather Viles is Reader in Geomorphology in the University of Oxford and Fellow of Worcester College. She has carried out research on many aspects of stone decay, including studies of the weathering of limestone at St Paul's Cathedral, London and investigations of the impact of traffic on building stone decay and soiling within the centre of Oxford.

Chris Wood is a senior architectural conservator in the Building Conservation & Research Team at English Heritage. He is responsible for managing several research projects and providing technical advice on remedial treatments to deteriorating historic fabric. He managed the English Heritage Master Classes at Fort Brockurst, (now at West Dean College) which specialised in the 'hands-on' repair of ancient monuments. Previous experience was gained as a director of a private sector architectural practice specializing in building conservation, and as a conservation officer in local authority planning departments.

Chris Groves is Director of the Hoffman Environmental Research Institute. He is a karst scientist involved in research on the Mammoth Cave karst system in Kentucky as well as a range of international projects. Much of his work revolves around developing and testing mathematical models to study karst aquifer evolution.

Part II

Development and case studies

Conserving fractured and detaching stone tracery
Developing a technique for the stabilization and consolidation of fire-damaged tracery at The Church of Holy Cross Temple, Bristol

Chris Wood *
English Heritage, 23 Savile Row, London W1S 2ET, UK
Colin Burns
14 St Cuthberts Lane, Locks Heath, Southampton SO31 6QR, UK

Abstract

The traditional approach to repairing badly fractured structural masonry is to replace the affected material with new replica stone. The wider use of resins in the last part of the twentieth century allowed the possibility of *in situ* repair but this had been tried unsuccessfully in the 1970s on the tracery at Bristol Temple Church. Nonetheless, English Heritage was keen to see if a refined method could achieve a successful result and be used as the basis for consolidating all the fractured tracery.

Key words

Conservation treatment, fractured masonry, resin grouts, wire reinforcement, Bristol Temple Church

INTRODUCTION

This paper describes unique remedial works to consolidate fire-shattered medieval window tracery in the bombed, ruined Church of Holy Cross Temple, Bristol (known locally as Temple Church, Bristol and referred to as such in this text). The work was to have two objectives: to pioneer and develop the technique, and to train the team who were expected to carry out the repairs.

In 1990 the English Heritage architect in charge of the ruins of Temple Church in Bristol described the tracery of the aisled windows as being 'in a dangerous condition held in place by obtrusive weldmesh grilles' (English Heritage 1990). This protection was inserted to prevent pieces of masonry falling onto visitors below. Clearly the condition of parts of the structure had deteriorated since its stabilization following Second World War bombing. The problem was that a very delicate repair method was needed, if replacement was to be avoided, since English Heritage's ethical stance ('preserve as found') required the most conservative treatment possible at reasonable cost. The technical difficulty was compounded by the failure of repairs carried out in the 1970s.

The proposal in 1990 was to repair the wall-heads and devise a method for consolidating and repairing the heavily fractured and detaching tracery below. The site was in the guardianship of English Heritage and at that time most repairs were carried out by the directly employed labour team (DEL), so the opportunity was taken to develop the repair methods with the team members and train them in the new techniques.

* author for correspondence

Clearly other works were also needed to minimize further damage by assisting the disposal of rainwater from the surrounding masonry and minimizing water ingress from the wall head above. Water was running down from the wall head parapet walk which was formed around a concealed concrete beam running the length of the building. A contract to weather the wall-head in 1991–2 was to be let, once services around the building had been investigated.

A number of principles were established during the planning of the remedial treatments. The primary objective was to design a process that would require the minimum amount of intervention, while at the same time conserving the tracery 'as found' (ie in the condition left immediately after the 1941 bombing and fires). Lessons would also need to be learned from the failure of the earlier technique carried out in the 1970s which had resulted in the loss of precious detailing. The repairs would also need to safeguard the structural integrity of the walls and ensure that the ruins would be safe and secure, as they would again be open to the public.

It was clear that a potentially lengthy and complex programme of works would be required. Therefore it was decided to pioneer the development of one refined and tested method of repair. Two training exercises were planned in order to develop techniques of stone stitching, grouting and consolidating surfaces. Before any work was to take place a detailed condition survey of all the tracery would have to be executed in order to help with the detailed planning of the trials.

While this survey was being carried out, an exercise in stone stitching and consolidation was organised at another English Heritage guardianship site at Minster Lovell Hall in Oxfordshire (OS Map 164; SP 324114). This offered an early opportunity to pioneer some of the techniques that would eventually be used on the tracery at Temple Church.

THE HISTORY OF TEMPLE CHURCH

Temple Church was one of Bristol's principal architectural glories (Fig 1). Founded by the Knights Templars in *c* 1140, the site was transferred to the Knights Hospitallers of St John after the ruthless suppression of the Templars in 1308. The church was then wholly rebuilt during the fourteenth and fifteenth centuries, largely funded by the profits from the thriving local wool trade. The Weaver's

Figure 1. The Church of the Holy Cross Temple, Bristol, taken in 1995 showing the south wall of the bombed nave.

Chapel (converted into a synagogue in 1786) dated from 1392, and the unusual hexagonal porch from the late fourteenth century, while the upper storeys of the famous 'leaning tower' (Fig 2) whose parapet overhung the base by 1.5 m (5 feet) by 1939, were rebuilt in 1460.

In 1941, however, much of central Bristol was obliterated during the notorious series of 'Baedeker' Nazi bombing raids, directed at British cities of historic rather than military importance (Fig 3). Temple Church itself was reduced to a shell by incendiary bombs, whose searing effect on the local limestone walls is still shockingly evident (Fig 4).

Immediate stabilizing repairs were carried out after the war, but for the next thirty years the church remained in very much the same ruined condition. In December 1958 it passed into the Guardianship of the Ministry of Works, although still owned by the Bristol Diocesan Board of Finance. Various works of structural consolidation were carried out to make the structure safe. The condition of the tracery was clearly of concern but as this did not threaten the stability of the ruin and further damage and loss could be controlled by the weldmesh grill, masonry repairs were not put in hand until the 1970s.

In an attempt to prevent further decay of the tracery, fibreglass[1] pins set in epoxide resin were inserted into the masonry in an effort to bind the fragmenting stones together. The principle was a good one, but the execution was flawed: not only were the pins oversized, they were also implanted in locations which were already under considerable structural stress. The result was that the introduction of the pins hastened, rather than halted, the shattering of the stonework (Fig 5). Large format plastic pins have high coefficients of thermal expansion, eleven times that of masonry (Fidler 1982) and resultant thermal movement forces become disruptive.

THE STONE

A sample of stone (a prism shape measuring roughly 75 x 38 x 50 mm (3 x 1.5 x 2 in) was analysed by English Heritage's consultant geologist (Dimes 1990). A visual inspection was made, and with the aid of a x10 lens and stereoscopic microscope and simple chemical and physi-

Figure 2. The famous tower, a Bristol landmark, whose parapet overhangs the base by 1.5 m (5 feet).

Figure 3. The effect of the incendiary bombs.

cal tests it was also compared with material from known localities and geological horizons. The specimen was described as a cream-coloured (between 5Y 7/2 'yellowish gray' and 5Y 8/4 'grayish yellow' on the Rock-Colour Chart), [2] granular, highly calcareous, with some small amount of comminuted [3] fossil matter.

Freshly broken surfaces had a saccarcidal ('sugary') appearance. The specimen was classified as a limestone and in general appearance and from close comparison with known material it was determined to be an example of Dundry stone.[4] This determination was felt to be sufficiently secure to not require a thin-section for microscopical study or petrographical description.

ETHICS

English Heritage's approach to repair is set out in *The Repair of Historic Buildings* (Brereton 1992) and accords with principal international conservation charters (ICOMOS 1966). In this instance the aim was to stabilize the condition of the masonry to prevent further loss of fabric and carry out works that would minimize the harm caused by the normal agents of decay.

The church and the traumatic events of 1941 are very important to Bristol's history and so the principle of 'repair as found' would mean securing the fragmented masonry in a way that still showed the fractured and

Figure 4. Fractured reveals showing the distinctive 'fire-reddening' which is caused by excessive heat on the iron-rich minerals within the stone. See Colour Plate 25.

Figure 5. The repairs in the 1970s have led to the loss of nosings, leaving them standing proud, the opposite effect to that intended.

Figure 6. The fractured and damaged arches of Minster Lovell Hall, Oxfordshire, viewed from below, showing the reduced bearing.

reddened stonework in its ruinous state. Restoration or repair with new stone was not an option at this stage, even if appropriate material could be found. Taking the damaged tracery down and rebuilding it using original damaged blocks of stone might have been technically feasible (although many blocks were sheared and impossible to re-use), but again this would have destroyed the historical authenticity of this part of the building.

While the selected method of repair must allow the fabric to continue to function by supporting loads and accommodating movement and moisture transpiration, two questions remained: how visually obvious should the repairs be, and could they be made reversible?

Appearance was an important issue. The need to show fractures and discolouration of the stone was an agreed objective. Visitors, though, would not see detail close to, unless a scaffold or ladder was in place. It was agreed that repairs should not be obvious from the ground, but that an indication of the works should be evident with close inspection.

Reversibility is a fundamental aim of conservative repair. The justification for this is that more benign or effective repair methods and materials may become available in the future. In this instance it was held to be a desirable objective, but the overriding need was to achieve structural stability which suggested that it might be difficult to remove the work (if ever required in future) without jeopardising the fragile masonry.

DEVISING THE REPAIR METHOD AT MINSTER LOVELL

The Research and Technical Advisory Service (RTAS) of English Heritage (now renamed the Building Conservation & Research Team) received a request for technical advice on repair methods from the regional English Heritage team responsible for the remains of Minster Lovell Hall, Oxfordshire. Initially the request was for advice on mortar specification for the repointing, wall capping, plaster filleting and grouting of the internal faces of the Great Hall, but during the inspection it was clear that there was a larger problem. This revolved around two high-level, four-centred limestone window heads (internal) where one of the arches was scheduled for replacement.

The problem seemed to be due to the internal arch stones being face-bedded, which during the many years of exposure to the elements and constant rainwater run-off from the broken wall head above, had resulted in loss of fabric along the bedding planes of two adjoining arch stones (Fig 6). Fabric had been lost from the front face of one voussoir, and the adjoining voussoir had losses from the back face, resulting in an approximate loss of 50% of the joints' bearing area. Stress fractures were also present and the areas of erosion had been exacerbated by the concentrated rainwater runoff from the broken wall head 0.5 m (18 in) above.

It was agreed that a programme of repairs would be carried out by English Heritage craftsmen under the direction of the RTAS which would seek to ensure that the arch was retained in position. The four-day live-site training exercise was successfully carried out in September 1990.

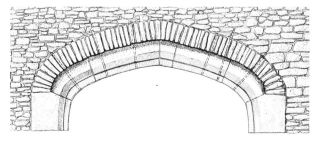

Figure 7. Method for securing voussoirs to the relieving arch using threaded stainless steel rods shown as broken lines (drawing by Judith Dobie, English Heritage).

Figure 8. Drilling through the voussoir beyond the relieving arch above.

Figure 9. Thorough flushing of all drilled holes with water prior to grouting.

Figure 10. Grout being injected up into the hole until it reaches the height of the hole in the face.

Figure 11. 6 mm threaded stainless steel rods being pushed gently through the clay plug and into position.

The objectives of the consolidation and conservation techniques undertaken on this arch were to provide mechanical support to the voussoirs. This was achieved by tying individual voussoirs to the relieving arch above. In this way, any further erosion at this critical joint would not result in collapse, but the transference of load to the relieving arch above. This was done by drilling two 10 mm (c $^3/_8$ in) diameter holes from the intrados (soffit) of each archstone, through and 50 mm (2 in) beyond the relieving arch over (Figs 7 and 8). Holes were then drilled at an inclined angle from joints in the masonry face above the relieving arch, to connect with the top of these holes. The holes were then thoroughly flushed with water (Fig 9) to remove slurry and debris, and also to thoroughly wet the internal core and masonry surrounding the holes, so as to limit subsequent suction upon grouting.

A lime-based formula was then injected using plastic syringes through the drilled holes at the intrados and this continued until the grout reached the height of holes in the face (Fig 10). This ensured that voids and fissures in the core and stonework leading from the holes were filled, a necessary precaution to prevent the resins, which would follow, from entering into such voids. When the grout had stabilized, the holes were re-drilled to the same diameter, cleaned out using bottlebrushes and water and allowed to dry.

Clay cups were formed around the tops of the holes at the face. Resin was injected through clay plugs at the bottom of the holes, continuing until it flowed into the clay cups at the top. The injection nozzle was then removed from the bottom hole, at the same time squeezing the clay plug to seal it and prevent any resin leakage.

Threaded $^1/_4$ inch (6 mm) diameter stainless steel bars, pre-cut to length, were then pushed at a gentle rate through the clay plug and unset resin into position within the drilled hole (Fig 11). During this operation, displaced resin was forced upwards into the clay cup. Here it acted as a reservoir to take up any further absorption of resin by the surrounding lime grout lining or masonry. As the end of the stainless bar passed through the clay plug, the clay was pushed into the hole behind the bar to a depth of 25 mm (1 in). This provided adequate cover and retained the bar in position until the resin began to cure, at which stage the clay and resin were removed from the ends of the holes. These were later filled with a lime-based mortar plug matching the stone. All fractures in the arch stones were prepared and filled with lime-based grout and all surfaces were given a lime-based shelter coat.[5]

This successful training exercise was used as the model for Temple Church, Bristol.

TEMPLE CHURCH, BRISTOL

Condition of tracery windows

There was very little evidence of lichens and vegetation on the tracery so it was clear that they were heavily

Figure 12. Fracturing along the line of the bedding planes caused by rapid cooling of stone following the fire. Frost has exacerbated these cracks over the years. See Colour Plate 26.

Figure 13. A photogrammetric drawing which was used as the basis for the detailed condition survey and shows the stones that were conserved during the trials (drawing by Judith Dobie, English Heritage).

fractured and fire-reddened over much of their surface, particularly at higher levels of the windows (Fig 4). Fracturing of stonework to this degree was caused by extreme and rapid changes in temperature caused by the heat of the fire and the application of cold water by the fire-fighters. The dramatic changes in temperature caused micro-fractures by setting up stresses which led to ruptures within limestone along the line and across bedding planes within the stone (Fig 12).

Permanent fire-reddening evident at the church was due to the change in the iron-rich minerals of the stone, which changed colour as water was driven off. The temperature at which this change took place is commonly quoted at about 500 °C (932 °F). Frost action widened these fractures over the years. The widths of fractures now varies through the stonework ranging from hairline fissures, up to 3 mm ($\frac{1}{8}$ in) generally and to as much as 8mm–12mm ($\frac{1}{4}$–$\frac{1}{2}$ in) in exceptional cases.

The initial survey and inspection of the window in the worst condition (Figs 13 and 14) confirmed that the most severe fracturing of the tracery stonework had occurred at higher levels and that these fractures ran in many directions and in a random manner.

Some shallow scales which carried surface detail had been lost many years earlier. Other areas, nearer the apex in particular, appeared soft and friable. On closer inspection these were found to be firm, but brittle to the touch. It was thought that this could be the result of lime consolidation caused by many years of rainwater percolation though the broken wall head and wall core, one metre or so above the apex of the window arch. This

Figure 14. The trials were carried out on the tracery on the window in the SE corner of the church, facing the tower scaffold.

Figure 15. The large ceramic dowels are evident just behind the nosing to the tracery which has now gone. See Colour Plate 27.

water would have carried calcium salts from the wall core which then formed a consolidating layer of calcium carbonate upon evaporation at the surfaces of the stone. This process of 'lime-watering' would have meant calcium hydroxide being carried in solution which would have crystallised to become calcium carbonate once more.

Although the masonry was very badly fractured the pieces were generally interlocked and, for the majority, unlikely to become dislodged. However they would still need to be secured because some of the fractures had split stones right down the middle and a single touch could result in further loss. Temple Church is also open to visitors so for reasons of safety it was essential to secure loose masonry that could become a potential danger. Taking the tracery window as a whole, it was concluded that any loss of fabric could have been extensive, progressive and potentially catastrophic.

During the initial site inspection it was apparent that earlier 'repairs' had been carried out in a number of different locations. This work was principally the stitching of fractured vertical nosings to mullions and to certain areas of tracery (Fig 15). Generally it appeared that stitching had been carried out to single large fractures and not to the very complex fractured stones that were to be consolidated in the current exercise.

Fifteen years earlier, a pioneering method had been devised by John Ashurst, then Principal Architect at the Directorate of Ancient Monuments and Historic Buildings, Department of Environment, with Renofors, a firm marketing proprietary materials and services [6] and together they had carried out the trial works. Renofors used their own brand of resins for anchoring glass fibre dowels across the fractured stonework. These dowels varied in diameter, the largest being approximately 15 mm ($^3/_4$ in). In addition, the distance between dowels, between 450 mm (18 in) and 600 mm (24 in) seemed to be excessive. The failure of these trials was ultimately due to the difference in thermal movement between the slender detached nosing and the mullion stone itself. The nosings, when restricted from such movement by the larger glass fibre dowels, developed stresses around the area of the dowel resulting in their complete detachment (Fig 16). The system seemed to work when the pins were inserted vertically down the length of a mullion but failures across the nosings were also attributable to the lack of dovetailing across the joints; another consequence of the very

Figure 16. A more dramatic example of the failure of this early repair where the dowels now stand proud and the nosing has become completely detached. See Colour Plate 28.

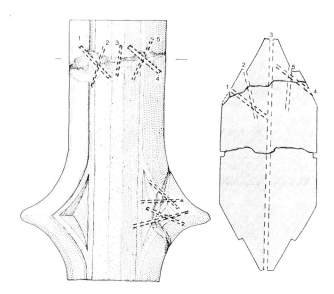

Figure 17. Sketch of a section and plan showing the principle of dovetailing the pins across the fracture and bedded into thicker parts of the stone (drawing by Judith Dobie, English Heritage).

wide centres (Fig 17 for dovetailing). Where smaller diameter dowels had been employed, with closer spacings, the repair had been successful.

The repair strategy

Devising the repair strategy began with the fundamental principle to 'conserve as found'. This meant repairing each stone but acknowledging that it should reflect the traumas of the war and still show discolouration and fracturing.

The first requirement was simply to secure together the individual fractured pieces of stone which formed a tracery block (each block being carved from a single piece of stone). This was essential in order to reintroduce structural integrity to the tracery window as a whole. The fractured blocks still had enough compression and shear resistance to prevent total collapse of the tracery, but lateral cohesion was undermined by the many vertical fissures in the stones. The repair strategy would have to overcome this problem.

It was important to ensure that repairs were confined to the individual tracery blocks and did not cross any joints within the tracery window itself or between the tracery and the first order of the surrounding arch. This would ensure that the masonry could continue to accept minor thermal or seasonal movement in the mortar joints, where such stresses are traditionally accommodated, without inducing or focusing stress at or around such points.

The next stage was to deal with the shock fractures within each tracery unit. These fractures varied in width from a few microns up to 10 mm ($^1/_2$ in). Some fractures were relatively clean and clear, others contained stone particles and debris. These needed to be thoroughly cleared and cleaned in order to allow free passage and a full filling of voids. They would also have to be filled with a material that would prevent rainwater collection within the fractures, which could lead to further jacking, wedging and widening during freezing conditions. The filler would have to provide free passage of moisture, both vapour and liquid, between the fractured stone pieces in order to allow natural moisture transference or breathing within the consolidated block.

Photographs and drawings were taken of the failed repair technique, which were mostly in the southern arcade, and of the fractured tracery window to be consolidated in the north elevation. These would be used to develop and determine the team's approach, along with notes made during the site inspection.

It was agreed that the usual form of pinning across fractures would be adopted, but that many small pins should be used to connect one fractured piece of stone to another. The pins should also be fitted at varying angles across fractures to provide a dovetail, which would lock one piece of stone to another (Fig 17). Unusually the pins would consist of two strands of thin copper wire. These would be twisted by hand or with cordless electric drills, restrained at the opposite end to form a spiral.[7] This would stiffen an otherwise thin and flexible wire as well as providing a key for the resin filler, which would hold it in position. Using two wires instead of one of a thicker gauge would improve the key. Several gauges of wire were used depending upon the strength required for the size of fragment to be pinned. Selection was based on experience and intuition, not calculation.

The holes drilled to receive the spiralled copper would be used for flushing out all fractures. This would mean that the debris within fractures would be washed from the heart of fractured stone, out to the surface which would be far more thorough than attempting to flush the fractures from the face inwards. After flushing, a lime-based grout would be introduced into these fractures through the holes drilled for the pins, following the success of this method at Minster Lovell. In this way, the grout would spread from the inside of the fracture, within the heart of the stone, outwards and upwards, thus ensuring all voids were completely filled. Once the grout had hardened holes were redrilled and the resin could be injected.

A suitable resin would have to be found to secure the pins in position. It would need to be water tolerant, as it would be placed within moist stonework. This was inevitable because of the limited time allowed for the training period and the wet processes involved in clearing and grouting the fractures. The problem of suction was not thought to be an issue, the objective being to find a resin that would effectively adhere to damp stone.

The problem of inflexible materials was a major issue so a consultant specializing in resins was asked to provide recommendations for appropriate formulations.[8] Resins traditionally set very hard to become rigid and incapable of accommodating movement. Stonework and mortar continue to move slightly through the seasons dependent on thermal and moisture capacitance. In any repair of historic material, it is essential to maintain flexibility, otherwise the stone is likely to come under considerable stress. This typically results in cracking at the junction of

Figure 18. Temporary support to prevent further detachment of fabric. Note the insertion of cotton wool behind the tape to prevent leakage during the grouting process.

stone and resin which over time could result in water ingress and damage by frost.

The selected resin would need to be applied by pressure from a hypodermic syringe with a 1 mm (0.04 in) diameter needle. It would therefore need to have a very low viscocity ie by being relatively thin to ensure an easy and continuous flow.

Possible contamination of the stone surfaces through accidental spillage was another concern with the use of resin because it can run easily and get into the pores of stone. Such is the adhesion once set that removal requires mechanical methods which invariably lead to the loss of historic fabric. It was agreed that the stones should be protected from this danger. As the surfaces were generally in a sound condition, it was decided to brush on a liquid latex rubber compound as a barrier, which, following the resin injection process, could be removed without damage to the surfaces involved.

Once all the materials and equipment had been prepared, two days were set aside for the site proving trials, in preparation for the live-site-training course the following week. In order to test fully the proposed methods and materials, the stones selected for the exercise were those in the worst condition. This would also be important for later scrutiny. The trials, though limited by time, proved to be successful and a programme for a four-day live-site training course for the DEL was prepared, with a list of materials and equipment that would be needed.

The procedure

The poor condition of some of the selected stones meant that gaffer tape (reinforced plastic adhesive tape) had to be wound around them to provide temporary support to prevent collapse during the works (Fig 18). On occasions, holes that would be later used for the spiralled copper had to be carefully drilled from one face to the other in order to accommodate such support (Fig 19).

Very careful consideration had to be given to every hole drilled. Observation of the direction that fractures appeared to take through the stone was critical in determining the direction and orientation of the holes that

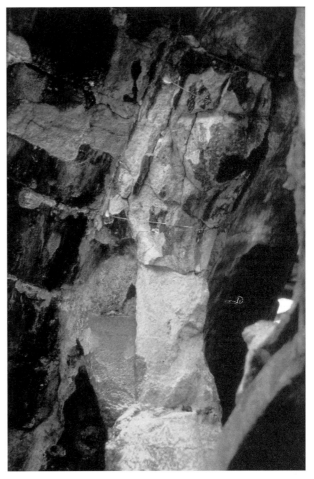

Figure 19. Holes had to be drilled to allow thin copper wire to be threaded through and around the stone to prevent detachment during the repair. See Colour Plate 29.

would accommodate the pins. To help determine the angles, thin wire (0.5 mm diameter [0.02 in]) was passed into the fractures from the face. Drilling confirmed the required angle of the first hole. Entering even hairline cracks concealed at the surface could be felt when drilling with a light pressure. Drilling generally went in perpendicular to the fracture but also at slight angle to provide a dovetail when the copper pins were inserted (Fig 20).

The types of drills selected were either hand or cordless electric non-percussive, with variable speeds. Standard tungsten masonry drill bits of the requisite size were not

Figure 20. Drilling square into the fracture and at a slight angle to accommodate the dovetail.

Figure 21. Using hand-held drills showing the requisite length of the cobalt bit.

available. The requirement was for a small diameter (3 mm [¹⁄₈ in]) and long length (at least 50 mm [2 in] beyond the fracture), the longest tungsten 3 mm (¹⁄₈ in) bit being 75 mm (3 in). Cobalt drill bits of appropriate size (Fig 21) were acquired from a specialist engineering tool supplier.[9]

The drilling of long, narrow holes required care and skill, if fragile masonry was to be drilled cleanly and accurately. Continuous drilling in moist stone could lead to a build-up of heat and turn the damp stone powder into a paste within the drilled hole. This could dry and become firm enough to stop the rotation of a drill, or, at worst, fracture or dislodge the piece being drilled. The speed of drilling varied and the drill was continually removed from the holes to enable the debris to be cleared. Each pass with the drill added about 10 mm (¹⁄₂ in) depth to the hole.

Remaining holes across the fracture could now be determined, ensuring that:

- holes were drilled a minimum of 25 mm (1 in) beyond the fracture
- the chosen configuration would evenly spread the load by making sure that the pins were concentrated in the thickest parts of each stone
- all holes were drilled at differing angles to each other, but wherever possible as close to 90° to the fracture line as possible, in order to provide a locking key once the pins were in position
- holes were not drilled into the narrow areas of fractured stone that ran to a feathered edge
- no hole came closer than 25 mm (1 in) to the face of any stone beyond the fracture.

Unanticipated fractures were occasionally encountered within the heart of the stone, despite following guidelines. When this occurred, the drill hole would be continued past the newly-found fracture in order to provide maximum support, on the proviso that there was sufficient stone to do this. When all drilling across fractures was completed on an individual stone, flushing would commence to wash debris from all fractures and drilled holes.

A gasket, formed from cotton wool or rubber 'O' rings, was fitted to the mouth of the hole and a small pressure spray was used to provide a continuous flow of water. This inevitably led to a great deal of spillage so the surrounding stones were covered with hessian and plastic sheets to avoid staining and ensure clean working. Heavier gauged copper wire twisted into a spiral was used like a bottle brush, to agitate against the sides of the hole and the resulting slurry was flushed out once more.

Pre-wetting

The grouting procedure began with the introduction of a pre-wetting solution, referred to as 'FSA' (flushing solution A). This comprised four parts water to one part alcohol (surgical spirit from a chemist). The purpose of this was to lower the surface tension of the water, which assisted the wetting of dry, dusty surfaces within the fractures. The solution was injected via the drill holes, using 50 cc hypodermic syringes (from laboratory suppliers) fitted with flexible rubber tubing. The tube was passed to the backs of the holes and sealed at the front with cotton wool, so as to soak up any excess and avoid spillages. The injected solution was thereby forced into and through the interconnecting fractures and drilled holes. Any escape routes were temporarily blocked using modelling clay or cotton wool. In this way the solution was encouraged to seek out and wet all surfaces, including hairline fractures.

Flushing and wetting

This was immediately followed by flushing with clean water; introduced by pressurised garden sprays held to the outside of the drilled holes with gaskets formed as before to prevent leakage.

During this process, small bottle brushes were run in and out of the drilled holes with water, to remove the slurry created during drilling. By damming the water

Figure 22. Flushing solution B.

within the stone, a reservoir of moisture was formed which would help to control suction, ie the de-watering of the lime-based grout, which was to follow.

Preparing fractures for grouting
Following the flushing and immediately prior to grouting, a second solution was injected, 'FSB' (flushing solution B, Fig 22). This solution comprised nine parts water to one part acrylic emulsion. When introduced prior to grouting FSB bound and strengthened friable surfaces and lined voids, which slowed the de-watering of grouts placed within them. This was injected at or near to the bottom of the stone, to a height of approximately 20–30 mm (¾ – 1¼ in). As the lime grout was injected, the FSB, being the lighter, was displaced and floated over the top of the grout. This was evident when the FSB reached the outside of the stone at fractures and drilled holes. In this manner, the FSB lined the surfaces of the voids immediately in front of the grout. As the FSB was used up, it was replaced at various heights through the stone by syringe injection via fractures or drilled holes. Immediately this was done, the lime-based grout was introduced (Table 1).

Table 1. Grout formulation.

Grout formulation	
Mature lime putty [minimum three months old]	1 part
HTI powder	1/4 part
Acrylic emulsion	1/10 part
Sodium gluconate / water [1:9]	1/100 part
Water	see below

The prepared mix using these proportions measured by volume was based on similar compositions developed for small-scale consolidation of fine fractures in wall paintings, by conservators working at ICCROM. [10] Mature lime putty [11] was specified because the longer it is left covered, the more water is taken up and the greater the proportion of lime to water. This was important in minimising the amount of cracking that could occur once the grout hardened. HTI powder provided some bulking out but it is also a pozzolan which provided a set to the mix, and in order to do this, it needed to be fresh. The acrylic emulsion slowed down the de-watering by retaining water which reduced shrinkage, as well as contributing some adhesion. The sodium gluconate was used to aid the flow of the grout.

Water content of the grouts varied, depending on the width of fracture or void to be filled. A general guide was followed, where fractures from hairline up to 2 mm ($c\ 1/8$ in) required two parts water, and fractures from 2 mm to 10 mm ($c\ 1/8$ in to $c\ 3/8$ in) required $1/2$ part water.

The variation in water content was to allow the grout incorporating two parts water to penetrate the finest of fractures, but would be unsuitable for any over 2 mm ($c\ 1/8$ in) range, because of the unacceptable shrinkage that would occur.

Water contents were intended as a guide only. A firm specification could not be given owing to a number of variable factors. The water content of lime putties can vary from one batch to another. The thoroughness of the wetting/flushing process and the porosity of the fabric itself will also have an influence on the amount of absorption in the stone. The final decision had to be made on site and was largely determined by the particular circumstances of each stone and practical experience of the materials being used.

Mixing the grout
The measured quantities of lime putty and water were thoroughly mixed, as they needed to pass through a 1 mm needle (see below). Fresh HTI powder was pre-sieved and added, followed by the acrylic emulsion and sodium gluconate. The grout was again mixed using both hand and power whisks, then passed through a 150-micron sieve to ensure that powder residues would not cause blockages during the injection process.

Grout was placed into suitable and stable containers with lids, to ensure that it was not contaminated with any

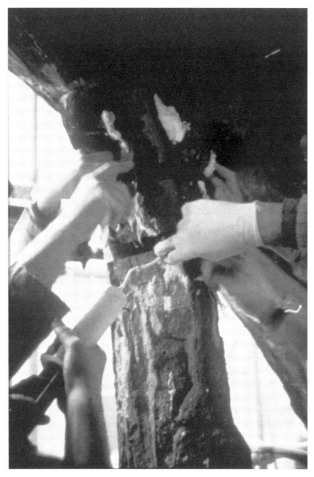

Figure 23. Grouting follows the flushing, here a plastic tube is being used for flexibility with plenty of volunteers to apply cotton wool if grout appeared unexpectedly.

debris during the process. The grout was then drawn into the plastic syringes, fitted with either clear, flexible plastic tubes or with 1 mm blunt needles, dependent on drilled hole sizes. Prior to the filling of syringes, the grout was re-mixed on every occasion to ensure that all particles were in suspension. Failure to do this would have resulted in the particles settling out and falling to the bottom of the container. This would only leave a thin limewater solution being introduced to the fractures, which would lack sufficient filler to bind it into a satisfactory grout. Unacceptable shrinkage would result in a significant loss of strength in the grout.

Figure 24. Clay cup acting as a reservoir of grout to fill cracks resulting from shrinkage when drying.

Figure 25. The grout has been effective at filling the larger cracks.

The syringes were inserted into the drilled holes and sealed around with cotton wool to avoid leakage; the aim being to get into the centre of the stone and force the grout from the heart of the masonry through the network of fractures and drilled holes to the face (Fig 23).

Grouting always began at or as near to the bottom of the stone as possible. This should produce an uninterrupted flow, upwards and through the network of voids. This method of working avoids air pockets, and any particles that may have remained following the flushing process are collected in suspension. It also avoids uncontrolled spillages. As the grout reached the face of the stone through these fractures or drilled holes, its flow was stemmed with cotton wool. The wool's absorbency acted to rapidly de-water and stabilize the grout near the surface with the added benefit of being self-supporting within minutes, provided grouting proceeded at the correct pace. The grout needs to flow gently but thoroughly throughout the network of voids so that it will give complete longitudinal filling. Too much haste can also lead to air pockets and voids not being completely filled. Exceptional pressure can lead to flakes blowing off and spillages.

This procedure was continued, changing injection points as required, until the grout reached the top of the voids. A clay cup was formed at the highest point of fractured masonry or drilled hole and was filled with

Figure 26. Applying the latex rubber to prevent spillage of resin affecting stonework.

Figure 27. Twisting the copper wire to form the spiral pins.

Figure 28. Clay cups with oversized twisted spiral pins in each hole. See Colour Plate 30.

grout to act as a reservoir, which compensated for any settlement or reduction in grout levels through water loss by suction in the masonry (Fig 24).

The water ratio of the grout was designed for the wider fractures (Fig 25); grout of this viscosity is unable to flow into the finer fissures that exist within the stone. A less viscous grout was then prepared and injected to the peripheral drilled holes or fractures, in order to fill these finer fissures. The procedure for injecting the thinner grout is the same as above, ie working from the bottom upwards.

The cotton wool used to dam the flow of grouts was left in place for several days to allow for a slow curing and prevent shrinkage. Had the cotton wool been removed straight away, rapid curing would have occurred on exposure to air, which would have caused de-watering and almost certainly resulted in cracks occurring. All holes for the pins (which had originally been used to introduce the grout) were re-drilled to remove the mix, and the holes cleaned with small bottle brushes to remove any residue and provide a suitable bond for the resin that would follow. At this stage the cotton wool was removed and the stonework given up to three coats of a brushable moulding latex rubber solution [12] to provide protection for the ashlar stone faces against any resin spillage (Fig 26).

Copper wires, twisted into a spiral using a drill and a pair of pliers (Fig 27) were now selected to suit the size of drilled holes, the largest of these being 1.2 mm (less than $1/16$ in) in diameter. The copper pins had to be cut to suit each hole owing to their different depths and every one was truncated by 10 mm ($c\,3/8$ in) to allow for adequate cover of lime mortar at the surface. As each one was cut for each hole they were left *in situ* (Fig 28). Before placing the pins within the holes, they were bent at the end to prevent them from falling through the downward sloping holes. For pins that were pushed upwards into holes, the bent end was wedged in the entrance to stop them falling out. Clay cups were then formed around each drilled hole in the stone, to act as both overflows and reservoirs during the resin grouting process.

The resin selected for use was Rutapox R1210 [13] because it was water tolerant and was to some degree, flexible. This meant that the grout would adhere to the walls of the drilled stone even though they were damp. The flexibility would allow the filled joint to accommodate minor movement expected in a masonry structure. As with most two-pack systems, extremely thorough mixing of the adhesive was essential for good performance. A whisk fitted to a cordless electric drill was used to mix the materials in a plastic container, thirty parts of hardener to one hundred parts of resin, measured by weight. Working time with this resin was between 15 to 20 minutes, before the initial cure began. The manufacturer's instructions were followed scrupulously, with just enough resin being prepared for a single stone.

Low viscocity resin was injected by syringe into the backs of the holes (Fig 29). The syringe was then withdrawn to ensure that the hole was completely filled to the stone surface and that no air pockets were formed.

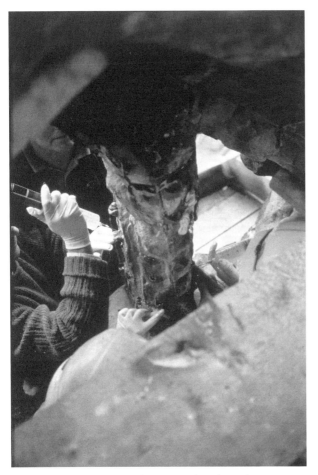

Figure 29. Low viscocity resin being injected on three sides.

Figure 31. The latex rubber protection being removed after the resin had cured. See Colour Plate 31.

Figure 30. Copper pin, now cut to size being carefully pushed into the resin-filled hole.

The copper pins which were oversized to accommodate the cranks had been withdrawn before the grouting, and straightened and cut to finish 10 mm (c $^3/_8$ in) from the face. These were then slowly pushed into resin-filled holes, so that air would not enter the resin in the spiralling of the pins (Fig 30). During insertion of the pin, resin was displaced into the clay overflow cup, which then acted as a reservoir to compensate for any lowering of the levels that occurred through absorption into the pores of the stone. As the resin began to cure and had formed a hard rubbery consistency, the clay cup was removed and resin excavated from the hole to a depth of 5–6 mm (c $^1/_4$ in), using either a pen-knife, scalpel or metal spatula.

When all holes had been treated in this way, the latex protection was peeled from the stone surfaces (Fig 31). At this stage, the lime grout within all but the very fine fractures was compressed at the surface with small pointing keys to ensure that no de-watering shrinkage had taken place. All fractures were raked back an average of 3–4 mm (c $^3/_{16}$ in) to provide a good key. Areas of stone surrounding the fractures were cleaned with a fine bristle brush and water to remove all traces of grout in preparation to receive a protective mortar fill.

Protective mortar

Mortar, consisting of one part of mature lime putty, two parts of washed sharp sand (1.18 mm down) [14] and one part of crushed limestone with 10% HTI powder, was

Figure 32. The 1990 work was 'overfinished' producing a rounded appearance rather than the dramatic, sharp, angular appearance more typical of fire-damaged fractures. See Colour Plate 32.

Figure 33. The 1991 work was far more successful at reproducing the original appearance of the damaged masonry. See Colour Plate 33.

Figure 34. Shelter coat being applied.

prepared, placed and fully compacted into all fractures using pointing keys or fingers protected with rubber surgical gloves. This mixture was designed to provide the flexibility and resilience required. The aggregate size was kept small because of the size of the opening, but nonetheless an even grading was specified to ensure that the lime binder was most effective. The crushed limestone was included to act as a porous particulate whose primary purpose was to add air and assist carbonation. A more robust mortar is produced if this is done slowly and the crushed limestone had the added advantage that it also provided a source of moisture which would slow down the eventual carbonation of the putty lime. The HTI powder provided an initial set before the longer carbonation process began.

The water content of the mortar was kept as low as possible, but consistent with high workability. During the trials at the first training course in October 1990, the mortar fill to the fractures was brought flush to the stone surfaces in general areas and was thickened within internal angles in order to provide maximum cover to the grout. The visual effect of this was to 'soften' the stone by giving it a rounded rather than sharp, angular appearance which is more typical of fractured stone (Fig 32).

In the second trial, following the second course in May 1991, the grout was raked back from the fractures to a greater depth. The lime-based mortar was prepared and placed as before, and provided the same degree of

Figure 35. Shelter coat after it had been 'bagged in' before being wiped down with damp sponges to bring through the colours and textures to match the existing

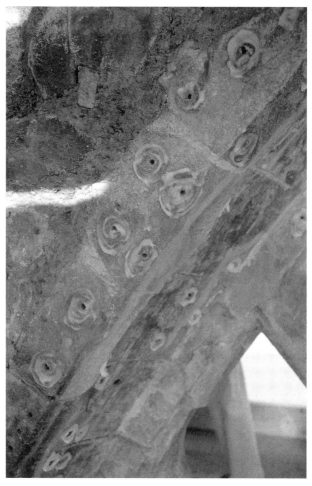

Figure 36. Nimbus Conservation Ltd used silicone around the holes to prevent resin spillage, rather than latex rubber. See Colour Plate 34.

protection, all sharp angles being maintained. This returned the angular, fractured appearance to the stones and visually was a great improvement as the consolidated stone now resembled the more dramatic appearance following the 1941 bombing (Fig 33).

Shelter coats

A lime shelter coat was applied to afford some protection to the work as water was still entering the stonework from above (Fig 34). The reddening of the stonework, caused by the fire, varied in depth within individual stones and in different parts of the windows. This resulted in a mottled appearance between the natural honey colour of the stone which was revealed where fragments had fallen away and the fire-reddened areas that remained. For this reason the shelter coats of lime putty, fine stone dust and casein were colour modified with natural earth pigments to match the existing colours of the stone. Yellow ochre was used for the natural stone colour, and Indian red and lamp black for the burnt stone appearance. Trial panels were prepared and left to dry until a satisfactory colour match was achieved.

There was still a reservoir of moisture within the stones from the first phase of repairs, but, even so, the surfaces of the stones were again sprayed with water to ensure that the shelter coats would not dry too rapidly. The face of the stone was wet but not glistening. The shelter coats had a thin cream consistency and were applied to the stones surfaces with a soft bristle brush.

This brushing process was important. Extensive brushing in varying directions, including small circular actions was essential in order to encourage the material into the pores and tiny interstices of the stone surfaces. When this had been completed, the shelter coat was given time to settle to a thick leathery consistency. The time taken was dependent upon the prevailing weather conditions. A soft bristle brush, which had been cut shorter to reduce and stiffen the bristles, was used to compact the shelter coat in the interstices of the surface, at the same time removing surplus material from high points of the stone which would preserve the sharp profiles of mouldings and other original features.

The result was to even out the surface textures, increasing water run-off (Fig 35). A final wiping of the surface with a damp sponge further enhanced its appearance by allowing the natural colour variations and texture of the stone to show through the shelter coat. The bland lifeless appearance so often encountered with shelter coating was thus avoided. The work was successfully completed and the DEL were confident about applying these techniques successfully to the whole of the church. Inspection of the shelter coat five years later found it to be still in good condition.

CONSOLIDATION OF THE CHURCH'S NORTH AISLE WALL AND TRACERY

Following the trials a contract for extensive works to the North Aisle wall tracery was let by competitive tender in 1994 because the DEL were being privatised. The successful tender was won by Nimbus Conservation Ltd, under the supervision of Philip Hughes, building surveyor,[15] who had produced a very comprehensive condition survey and specification. This was based on initial advice provided by members of the English Heritage South West Regional Team and BCRT who had been involved with the earlier trials.

These works were far more extensive and included capping the wall-head, cleaning, raking out cementitious joints and complete pointing of the walls.[16] The tracery repairs followed a similar strategy to the trials. A pozzolanic lime grout was used to provide initial adhesion of loose fragments so a quarter part of trass (a volcanic ash) was added to the aggregate mix, comprising one part of mature lime putty and half a part of silver sand. The trass would provide an initial set which was felt to be important as the work was to be carried out over winter. Its dark ash colour was also expected to blend with the original sooted lime mortar. The mix was varied dependent on the width of crack and accessibility. Several of the worst areas of detachment were dismantled, cleaned and refixed in a fine lime/trass mix. Dismantling was felt to be necessary because of the loss of stone between its main body and the fragment, or that its friability prevented a good bond, or the fragment had dropped to such an extent that it needed to be re-adhered.

Gravity grouting was carried out through various sized nozzles and syringes for the cracks that were not dismantled. This included grout being poured into fractures from above.

A similar process was followed using 1.2 mm twisted copper wire, with epoxy resin injected into the drilled and dusted holes. Sika 31 epoxy resin [17] was used as it provided a degree of flexibility and was gunned into the hole using nozzles with a 4 mm ($c\,^3/_{16}$ in) grade plastic hose attachment so that resin could be extruded from the bottom up. Holes were drilled to the required depth, the most frequent diameter being 5 mm ($c\,^1/_4$ in). However instead of using latex to protect the surrounding stonework, silicone mastic was used around the holes (Fig 36). Coloured mortar to match the stone was used to plug the holes. The red mortar comprised a mix of one part matured (one year old) lime putty to two parts Taunton red sand, and half a part of Guiting and Bath stone dust. A quarter part of trass was added to this coarse stuff. Yellow mortar comprised a similar mix with Dorset Ginger sand being used instead of the Taunton red. Stronger depths of tone were provided by using ivory black, terre verte and red ochre pigment.

Some limited stone replacement was undertaken as well as extensive surface patching of the stone using coloured mortars. Following the lime mortar repairs a colour-matched shelter coat using one part of mature lime putty to either four parts of fine red dust (three parts Taunton red and one of Guiting stonedust) or four and

Figure 37. The fully repaired window in 1996.

a half of yellow dust (two and a half parts Tetbury stonedust and two parts Hornton brown) and dry casein (between 10 and 12%) was worked into open pores and water traps in the stonework.

As the work was undertaken during winter, protective sheeting was used on the scaffold together with insulated blankets to provide protection from frost and rain. This was another reason why trass, a pozzolanic additive, was used in order to provide an initial set.

CONCLUSIONS

In order to evaluate the performance of the trials carried out in 1990 and 1991 it is necessary to consider the original aims of the project. These were:

- to clear and clean all the masonry fractures of dust and debris and fill them with a medium which would allow free and ready passage of moisture from one fractured section of stone to another
- to prevent the collection of free water within any fracture to avoid potential damage from freeze/thaw cycles
- to provide enough flexibility to accommodate thermal movement within the masonry unit
- to connect all of the fractured pieces which form a single masonry unit so as to reintroduce structural stability, without inducing harmful stresses to any part of the repaired unit

Figure 38. Detail of the 1991 repair taken eight years later, still looking satisfactory, although as expected there has been some loss of shelter coat. See Colour Plate 35.

- to provide a sympathetic lime-based mortar fill to the outer edges of fractures and any areas of repair
- to provide a suitably protective coating to the masonry unit which would enable rapid run-off of rainwater, would not interfere with the natural breathing of the stonework and would allow the texture and colour variation to remain visible.

The first close inspection of the work was carried out by BCRT in 1995. This included one member of the original team that devised the repair system and provided the on-site training. The condition of the stones four years on was encouraging and very gratifying, particularly as the tracery was still subject to a certain amount of water ingress from the wall-head above (Fig 36). This was subsequently repaired later that same year.

The last inspection of these works was undertaken by BCRT early in 1999 (Figs 37, 38 and 39). The repairs continue to perform without any visible sign of failure or distress, apart from the shelter coat which was showing signs of material loss. Re-application was therefore recommended within two years.[18]

Any new repair system, in common with many other conservative interventions, will require development and testing in order to refine and obtain the required results. Many factors will influence the success of a treatment, particularly when a number of different materials are brought together in a repair. We must understand the materials, in terms of their performance in differing and constantly changing environments, and, most importantly, how they will react with each other. Permeability, durability, expansion coefficients of the varying materials, exposure, climate, required surface

Figure 39. Looking up at the head of the window in 1999 which shows it to be in good condition.

textures and final appearance are some of the more obvious factors that must be taken into account when devising a repair strategy. It is a rare situation indeed when the first trial produces a solution that does not require some adjustments or that could not be improved upon through better material selection or technique.

It would appear, judging by the condition of the consolidation at Bristol Temple Church, that this exercise could be termed a success. However, nine years is a relatively short period of time to make such claims. There were a number of issues that caused concern to the team during the initial development work. These were, principally, the use of resin and copper wire and the choice of a lime-based grout for filling the voids. Resins have traditionally been regarded with scepticism, particularly for the conservation of limestone. This is because of resins' inflexibility and impervious qualities which creates problems with moisture transference. However the judicious use of the material here seems to have been successful. The obvious alternative today would be a hydraulic lime-based grout, but in 1990 there were few reliable sources available. It is also debatable whether hydraulic lime would have performed better.

Concern about the use of copper stemmed from the possibility of moisture bleeding producing a green copper sulphate stain. However the resin has prevented this possibility by effectively 'locking up' the copper. Stainless steel was the obvious alternative, but by comparison this is a hard and inflexible material. Liquid rubber was also considered but there was concern about how long it would take for the material to go hard and then lose its flexibility. Ceramic dowels were rejected because it would be difficult to obtain the small diameters required and in any event, such dowels would either be too brittle or too rigid.

The use of HTI as a pozzolan within the lime putty grout was again largely determined by the materials available at the time. Nowadays a feebly/moderately natural hydraulic lime would probably be chosen as the first option, as in early tests under English Heritage's research project AC 1: The Smeaton Project (Ashall et al, 1996) HTI powder was found to be quite vulnerable in its hydraulicity. We do not as yet have the depth of experience of any of these products to predict longer term performance. Clearly today there is more than one option. Experienced and thoughtful practitioners may well be successful whichever option is actually employed. Continuous evaluation of the performance of the trials (and the subsequent work by Nimbus) will continue to provide valuable information about the longer-term success and performance of this particular repair method.

ENDNOTES

1. Glass reinforced polyester (GRP) resin from the UK agents for Bêta International Trading (see product suppliers list below).
2. *The Rock-Colour Chart*, distributed by the Geological Society of America, is available in the UK at £17.50, from the Geological Society Publishing House, Unit 7, Brassmill Enterprise Centre, Brassmill Lane, Bath BA1 3JN; Tel: + 44 1225 445046; Fax: + 44 1225 442836.
3. Comminuted: the breakdown of a solid to a fine powder, usually by mechanical means such as grinding between glacier ice and rock.
4. Dundry stone was quarried from a hill at Dundry, Somerset, which was an outlier of Inferior Oolite, Jurassic in age, resting on the Lias. Here some c 5 m (15–16 feet) thick Dundry Freestone was found, of which about 2 m (6 feet) was worked as Dundry Stone, resting on about 12 m (40 feet) of Rag Beds. The stone was extensively used in and around Bristol and was also exported. It has been worked since at least Norman times, evidenced by the Norman gateway west of Bristol Cathedral and in 1300 Crown returns show that old quarries existed at the highest point of Dundry Hill, where there are still tracings of old workings. By 1990 the old quarries were being filled with rubbish which meant that authentic replacement stone would be impossible to source.
5. The materials and mixes used at Minster Lovell were similar to those used at Bristol and are described in more detail later in the text. Rutapox 1210 epoxide resin was used and the grout formulation was identical to that used at Bristol. The mortar mix comprised one part mature lime putty to one part crushed limestone to two parts of sharp sand (sieved to 1.18 mm down).
6. Renofors Ltd of Bolton, Lancashire, then retained the UK license for a Dutch patented system of glass fibre rod reinforced epoxy resin structural repair, utilized in timber engineering and for the consolidation of masonry. The system is still in wide use in the Low Countries. See product suppliers list below for Bêta International.
7. The term 'spiral' or 'pin' has been used to describe the twisted copper. It could more accurately be called a *helical* for single strands or *cabled* for two pieces twisted together.
8. The consultant is no longer contactable.
9. The company who mainly supplied the steel fabricating industry is no longer in business.
10. ICCROM is the International Centre for the Study of the Preservation and the Restoration of Cultural Property in Rome, at 13 via di San Michele, 00153 Rome, Italy. Tel: + 39 0658 5531. Website: www.iccrom.org. For details of the grouting technique see Ferragi et al 1984.
11. Mature lime putty is available from a number of suppliers, most of whom are listed in Tuetonico 1997. HTI powder was supplied by Steetley Refractories, Dudley, West Midlands who are no longer in business. Supplies can now be obtained from Rose of Jericho (see product suppliers list below).
12. A pre-vulcanised liquid latex was used from a 5 kg (11 lb) container, product code 404-110. See product suppliers list below for Alec Tiranti Ltd.
13. Rutapox 1210 is no longer made. The nearest equivalent which is very similar is EPIN07, produced and supplied by the same manufacturer, Krämie-Chemie (see product suppliers list below).
14. 1.18 mm refers to the mesh size of the standard sieves used for grading aggregates. Well-graded aggregates from 1.18 mm down would result in a similar proportion of sharp sands which would pass through mesh sizes of 1.18 mm, 600, 300 and 150 microns respectively. For more information on the selection, grading and availability of suitable sands for use in conservation, see Chapman & Fidler 2000.
15. Nimbus Conservation, Eastgate, Christchurch Street, Frome, Somerset BA11 1QD, UK; Tel: + 44 1373 474646; Fax: + 44 1373 474648; email: enquiries@nimbusconservation.com. Phillip Hughes Associates, Old Manor Stables, Tout Hill, Wincanton, Somerset BA9 9DL, UK; Tel: + 44 1963 824240; Fax: + 44 1963 824642.

16 Details taken from an unpublished report (Nimbus Conservation Ltd 1995).
17 Sika 31 is produced by Sika Ltd (see product suppliers list below) and can be obtained through most builder's merchants.
18 English Heritage research (Woolfit, this volume) has revealed little maintenance of lime shelter coating in the UK after its deployment. Coats need maintenance and replenishment every five to seven years.

PRODUCT SUPPLIERS

Bêta International Trading BV, PO Box 128, NL-4920 AC Made, Eerste Industrieweg 1, NL-4921 XJ Made, The Netherlands; Tel: + 31 162 672267; Fax: + 31 162 672277; email: renocon@wxs.nl; web: http://www.amsterdam.nl/bmz/conserduc

Krämie-Chemie, Theodore-Heusse-Str., 11-15, D-66130, Güdingen, Germany.

Rose of Jericho, St Blaise Ltd, Westhill Barn, Evershot, Dorchester DT2 0LD, UK; Tel: + 44 1935 83676; Fax: + 44 1935 83903; www.rose-of-jericho.demon.co.uk

Sika Ltd, Watchmead, Welwyn Garden City, Hertfordshire AL7 1BQ, UK; Tel: + 44 1707 394444; Fax + 44 1707 375593; email: sika@uk.sika.com; web: www.sika.com

Alec Tiranti Ltd, 70 High Street, Teale, Reading, Berkshire RG7 5AR, UK; Tel: + 44 118 930 2775; Fax: + 44 118 932 3487; email: enquiries@tiranti.co.uk

BIBLIOGRAPHY

Ashall G, Butlin R, Martin W and Teutonico J M, 1996 Development of lime mortar formulations for use in historic buildings. A report on the Smeaton Project, in Sjostrom C. (ed.) *Proceedings of the Seventh International Conference on the Durability of Building Materials and Components, Stockholm, Sweden, 19-23 May, 1996*, London, Spon, 352–360.

Brereton C, 1992 *The Repair of Historic Buildings: Advice on Principles and Methods*, London, English Heritage.

Chapman S and Fidler J (eds), 2000 *The English Heritage Directory of Building Sands & Aggregates*, Shaftesbury, Donhead Publishing.

Dimes F G, 1990 *Report on Stone used at Bristol Temple Church*, unpublished report for English Heritage.

English Heritage, 1990 *Condition Report: Bristol Temple Church*, unpublished architect's report.

Ferragi D, Forti M, Malliet G, Teutonico J M and Torraca G, 1984 *Injection Grouting of Mural Paintings and Mosaics*. Rome, ICCROM.

Fidler J, 1982 Glass-reinforced plastic facsimiles in building restoration, *Bulletin of the Association for Preservation Technology*, **XIV**:3, 21–26.

ICOMOS, 1966 *The Venice Charter, 1964*, Paris, The International Council on Monuments and Sites (ICOMOS).

Nimbus Conservation Ltd, 1995 *The Report for the Masonry Consolidation to the North Aisle Wall at Temple Church, Bristol*, unpublished report for English Heritage (SW Region) following completion of the works in October 1995.

Teutonico J M (ed.), 1997 *The English Heritage Directory of Building Limes*, Shaftesbury, Donhead Publishing.

ACKNOWLEDGEMENTS

The authors would like to thank Niall Morrissey (English Heritage South West Region) for his help with information on the previous and subsequent repairs to the ruins. All photographs are by the authors unless otherwise stated and are copyright of English Heritage.

AUTHOR BIOGRAPHIES

Chris Wood has worked in the Building Conservation & Research Team at English Heritage for the last seven years. He is responsible for co-ordinating research programmes, providing specialist advice on remedial treatments, running training and outreach programmes and leading national campaigns. He managed the English Heritage practical training centre at Fort Brockhurst, which offered specialist practical training concentrating on the repair of masonry structures. These courses are now being run at West Dean College, Sussex. Prior to joining English Heritage he was a director of an architectural practice specializing in the repair and refurbishment of historic buildings. This followed 12 years as a conservation officer with two local authorities.

Colin Burns is a stone mason with over 25 years experience working on sites for English Heritage and its predecessors. Colin joined the Building Conservation Team in 1988 providing specialist advice on masonry repairs and carrying out on-site training exercises. He was the Senior Training Officer at Fort Brockhurst where he helped to design and build the 'ruinettes' and walls that were used for the 'hands-on' elements within the Masterclass courses. He is now a consultant, but carries on this role at West Dean College, Sussex. He also provides 'on-site' training and technical advice on rectifying problems with historic masonry structures.

A solution for the stone repair of a cracked primary column at the Wellington Arch, Hyde Park Corner, London

Les Ayling
English Heritage, 23 Savile Row, London W1S 2ET, UK

Abstract

During the conservation project on the Wellington Arch, London, a sloping crack was discovered in one of the main columns. The repair was achieved with minimum intervention and disturbance to the entablature that it supported. A systematic appraisal of possible options culminated in a solution that only required the insertion of small fingers of stone, but which maintained the load on the column without the settlement problems associated with using large replacement stone with a new bedding.

Key words

Stone repair, stone column, chain saw, Portland Stone, minimum intervention-maximum retention

INTRODUCTION

In the mid eighteenth century, at a time when London was considered to end at Park Lane, Hyde Park Corner was considered an appropriate site for a grand western entrance to the city. A triumphal arch was suggested but nothing came of the designs.

In 1825–6, the architect Decimus Burton was asked to design new park entrances for Hyde Park and Green Park on this site, facing each other across Piccadilly. His design resulted in the screen that now stands at the south-east corner of Hyde Park and, directly opposite, the arch (Fig 1), which, according to Burton, was to be a new entrance to Buckingham Palace with side-gates to Constitution Hill and Grosvenor Place. Inside the Arch there were two Porters' Lodges, one for the royal gate and the other

Figure 1. The Wellington Arch, Hyde Park Corner, London, after conservation and restoration (photograph by Les Ayling, English Heritage).

for the public gates, which were to stand either side of the Arch. Burton's design for a royal gateway failed to be co-ordinated with John Nash's ideas for the remodelling of Buckingham Palace and it never achieved this status. The Arch was built and paid for by the Office of Woods and Forests and was finished in 1828 without much of the rich decoration, sculptures and the quadriga shown in Burton's design. It was variously referred to as 'Royal entrance at the top of Constitution Hill', 'Archway and Lodges at Hyde Park Corner at the Entrance of St James Park' or 'New entrance at Constitution Hill'.

In 1838 a Wellington Memorial Committee was set up to raise funds for a national memorial to the Duke of Wellington and in 1846 the Arch was topped with a colossal equestrian statue of the Duke. This statue was not well liked but remained until 1883 when, due to increasing road traffic in the area, the Arch was dismantled and re-erected on its present alignment on the axis of Constitution Hill. The Duke of Wellington statue was not replaced.[1] The Arch, now referred to as the Wellington Arch, eventually received a quadriga in 1912 when the sculptor Adrian Jones closely modelled the present bronze group on a small group called *Triumph* that he had exhibited at the Royal Academy in 1891.

Continuing traffic problems saw the London County Council construct the present large roundabout and underpass that the Duke of Wellington opened in 1962. The floors and roof in the northern side of the Arch were removed for the construction of a large ventilation shaft for this underpass. The south side of the Arch was used as a police-station, once said to be the smallest in London, until it was closed just prior to these works. The Wellington Arch, a Grade I listed building, has stood empty ever since.[2]

THE CRACK

In April 1999 English Heritage took over responsibility for the Wellington Arch, which was on the English Heritage *Register of Buildings at Risk*. The condition of the stonework, lead and asphalt roof coverings, hidden steel and iron support beams and bronze statue was suspect and a £1.5 million conservation and restoration project began to redress its decay. In October 1999 work started, directed by English Heritage's Major Projects Team and carried out by the main contractor, Mansell PLC. Responsibility for all structural engineering work was supplied by English Heritage's own team of specialist structural engineers.

Various structural defects, both stonework and ironwork, required attention during the conservation works. Scaffolding, covered with protective sheeting, was erected and the statue and Portland stone stonework were carefully cleaned and inspected to identify any further faults.

One fault, which came to light once the scaffold enabled very close inspection, was in one of the inner columns on the north-east corner, a crack running at an angle through a large portion of the column. On the west and east faces, the Arch has flat roofs surrounded by a parapet of large stone blocks which surmounts a decorative entablature supported on its outer edge by four large fluted columns, approximately 912 mm (3 ft) in diameter. The inner two columns form the support for a pair of cast-iron beams hidden within each of the entablatures that span the roadway running through the Arch. It was in the upper part of one of these primary support columns carrying the massive parapet, entablature and roof area that a sloping crack running around 60% of its circumference was discovered. Because of the angle of the crack the compression force from loads carried encouraged the column to have a propensity to shear sideways and, potentially, collapse.

Assessment

As there were no signs of movement, the column was obviously supporting its load by the interlock of the irregular crack line and by redistributing the loads down the uncracked section of the column. Any disturbance or vibration might cause the crack to extend further, thus reducing the remaining solid portion or fracturing the interlock. The crack could have been a natural fault line in the stone block or have been propagated as part of the quarrying, but the length of the crack made it difficult to see how it could have survived the dismantling and rebuilding of the Arch in 1883. The crack may have been caused by this disturbance, nevertheless, it was a structural fault to an essential support element that had to be addressed. Adjacent to the crack were indications of old stone repairs which may have been part of previous attempts at repairing this crack, so the area was surveyed for signs of any imbedded metal. The readings from metal detectors[3] only gave indications of a metal pin in the centre of the column at the bed joints between the shaft drums.

The crack, which is in the upper part of the column 9 m (30 ft) above the ground, runs down the northern side at approximately 45 degrees around the inner face and up at a similar angle on the southern face, thus appearing to form a slip plane (Fig 2). Any scheme of replacement or repair, therefore, must be capable of preventing any slip sideways while maintaining the load-carrying capacity, be conservation-friendly and involve minimal disturbance leaving minimal evidence of the intervention.

Figure 2. View of the crack on the north side of the column (photograph by Les Ayling, English Heritage). See Colour Plate 36.

Discussions ensued between team professionals and different ideas were examined, on paper at least. The different potential solutions were discussed with the contractor, as planning of the temporary works and any heavy lifting that would be required would have to be co-ordinated with the existing works.

Options considered

The process considered for dealing with this fault were:

- to replace the complete drum
- to remove the crack by cutting out large stepped blocks
- to inject the crack with resin
- to insert dowels
- to strap the halves together.

The contract included a stonework subcontractor, so it would have been seemingly straightforward to cut out the complete drum and replace it with new stone. While this approach would completely remove the crack it had its drawbacks. The access and support scaffolding were already in place and the whole Arch was covered with protective sheeting. Manoeuvring a large block of stone into position would have required strengthening of the access scaffolding and modifying the extensive support and bracing system to the upper part of the scaffolding, with considerable resultant costs and disruption to the construction programme.

Even if this approach had been accepted, other consequences of this solution would have been, for example, unwanted vertical movements of the entablature and its two cast-iron beams. With the complete drum removed and the entire load carried on a temporary support structure, the entablature would settle as this structure took up the strain. In addition, large areas of new mortar bedding and packing of the top surface of the stone would also lead to 'bedding in' and further, unacceptable, vertical movement.

Consideration was also given to avoiding the formation of an impervious barrier by the necessarily early-strength bedding mortar. This might interrupt the natural water percolation paths that occur during the wetting and drying processes within stonework and which could cause stone deterioration.

The stone carving on the Wellington Arch is exceptional and one of its main features, the loss of this amount of original stone was not acceptable. The choice of complete replacement is not regarded as conservation or as minimum intervention and so this approach was rejected.

The next option considered was to remove the crack by cutting it away, in stages, in three or four large adjacent stepped blocks.

The advantage of this method over the previous one of complete replacement was that more of the original stone would remain supporting the load in whatever way it had become accustomed to over time, until all the blocks had been replaced in turn. Movements due to disturbance would be less, as the bedding of each piece could be allowed to achieve full strength before cutting away the next block. This method still achieved the removal of the complete crack.

The disadvantage, however, was that although smaller pieces of stone could be used there was still a considerable amount of original carved stone to remove. The disturbance resulting due to the cutting operation on blocks of this size could break the fragile interface of the crack allowing unwanted settlement as described with the previous method. There would also be some difficulty in guaranteeing full grouting of the stone, deep within the column, especially the vertical faces and in any case this would again lead to the new stone being enclosed in impervious mortar.

While the bedding and settlement aspects were still not ideal the proposal was an improvement on the previous solution, but the resulting patchwork of new stone at such a prominent position was not considered to be the best way of conserving historic fabric. Although a better method, it was not acceptable. A simpler and neater solution was called for.

Doweling was the next proposal to come under scrutiny, the use of stainless steel rods inserted at right angles across the crack plane.

The stone column could be drilled with an array of holes into which stainless steel rods would be epoxide resin-grouted into place across the crack. To ensure they were placed at right angles, the insertion of some of the dowels would have entailed drilling a deep hole, some starting in the stone above the damaged one.

The use of dowels to mechanically tie pieces of stone together is widely practised. For a small amount of intervention they are, in many instances, very effective. For dowels to be effective in this situation, the following criteria must be satisfied.

- The dowels must be able to resist a predominately shearing force, a component of the vertical load, acting along the slip plane produced by the crack.
- The stone must be able to resist the local crushing induced by the dowel.
- There must be enough space between the dowels and from an edge to dissipate the forces.

Unfortunately, in this case, these criteria could not be satisfied.

The design calculations revealed that these stresses were too high for the Portland stone to resist and the total number of dowels required was prohibitive. There was not enough space available.

It had to be borne in mind that, because the crack was at 45 degrees, the extremities of the stone became a circular shaped wedge with little strength. It, therefore, did not lend itself to having dowel forces acting close to it.

There were other factors against this approach.

- The successful filling of the dowel holes with grout would also be critical to their load-carrying capacity but this could not easily be guaranteed in deep, small-diameter drillings.

- The extent by which a multitude of deep drillings could be damaging to the good stone above and away from the crack.

Dowelling with multiple rods was, therefore, dismissed as an option.

With this limitation in space for small diameter dowels, a check was also made utilising a single large diameter dowel, a plug of stainless steel rod of about 100 mm (4 in) diameter. This meant only one hole to drill and grout with one strong steel dowel.

However, this solution, too, would produce not only stresses greater than the stone could resist but also a very large circular hole through good stone, and a probable and unacceptable blemish on the flutes of the drum. It was similarly rejected.

Another solution that received an assessment was the injection of the crack with resin to 'stick' the two halves together.

This would appear to be a very simple remedy. It would require little or no temporary propping, and therefore would not disrupt other works and there would be practically no physical or aesthetic intervention.

Questions to be resolved with this method were:

- Would the resin penetrate the whole surface of the crack?
- Would the resin be strong enough to adhere to the crack surface?
- Would the crack have loose fine grains of stone dislodged during the break which would prevent the resin from forming sufficient bond to each half?
- While there are low viscosity epoxide and polyester resins that will percolate fine cracks, what was the extent of the crack within the column?
- Where should the injection points go?

Because of the depth and shape of the crack, injection from nipples placed along the surface line of the crack could not be seen to give full coverage of adhesive. The only way to attempt to achieve full coverage was to start injection from within the column and force the resin to the surface, expelling the air from the advancing resin. Drilling the holes for these injection tubes without knowing exactly where the internal line of the crack extended, and hence starting the grouting at the correct internal point, would be impossible.

Although the crack was clearly visible, it was nevertheless under compression forces without any discernible gap. At the time, the depth and exact internal direction of the crack was not known. The angle of the crack was the same on both sides of the column and, therefore, it was logical to suppose that the crack extended right through the column as a single plane. But it was only an assumption.

Apart from the difficulties of ensuring full resin coverage, pressure-grouting resin into the crack has other distinct disadvantages; the pressure of the injection would tend to force the crack apart and the resin itself might act as a lubricant to the slip plane mentioned before. Both of these would have been very undesirable in the circumstances.

Adhesives are formulated to give properties to suit many and various applications. Most research and information concerns their use in controlled factory-type situations where the design of the component and the adhesive type used can be altered to maximize the efficiency of the jointed product. In the conservation of stone structural elements these choices are much reduced, with variable environments and stone types variable and parts that cannot be the optimum shape, an important consideration for glued joints.

The strength of any glued joint will depend upon the strength of the adhesive and the degree to which it bonds to the components. Adhesion at the interface of adhesive and component occurs within a layer of molecular dimensions. It follows that very careful selection is required, choosing an adhesive for one particular property of a joint may render it useless when checked against other requirements.

The adhesive must have the ability to wet the surfaces of the crack and flow into all the irregularities so that a high degree of contact is achieved, and then solidify to form a lasting joint. Surface contaminants can greatly or even totally reduce any bond and any adhesive starvation within the joint will not allow design strengths of the joint to be reached.

An important consideration when adhesives are suggested as a method of repair is the design of the joint. The strength of the adhesive must be assessed not only under uniform loads, shear, tension and compression, but also peeling and cleavage forces, as these produce areas of high stress concentration. Peeling forces are not usually a consideration with structural repairs as at least one component must be flexible. Cleavage forces, on the other hand, may arise from offset tension force or applied moment, a situation that can occur if the joint is still subject to movement. This force concentrates stresses at one end of a joint. An adhesive should not be used in a situation where there should, in fact, be a permanent movement joint.

Adhesives for application to very porous faces often require pre-treatment with a surface sealer. This is obviously not possible unless the parts can be dismantled.

Low-viscosity adhesives suitable for penetrating into fine cracks have the disadvantage of being absorbed into porous stone and be liable to produce surface staining along the joint line.

When fillers and thickeners are added to produce thixotropic properties, which allow the adhesive to harden once left undisturbed, the adhesive may then have limited ability to penetrate very fine cracks.

The ideal condition for joint faces is to be free of dust and debris, and dry, although some adhesives will tolerate dampness and will work to encapsulate joint debris in its glue line. These may not satisfy other requirements.

The use of adhesives for the repair of structural load-bearing stone must be very carefully considered. The variable nature of the stone, the condition of the crack surfaces and the site conditions under which the adhesive

Figure 3. ICS® 823H SAW (ICS® Blount Europe).

is applied all tend to reduce the effectiveness of the joint strength.

The designer must be satisfied that the forces can be safely transferred from the constituent stone of one half, through the glue and back into the stone of the other half and do it without degradation in conditions of heat, cold, age or other in-service conditions.

There seemed little sense in committing funds to any kind of ultra-sonic or radar surveying techniques to determine the exact internal line of the crack, as it was becoming evident that the solution lay in more extensive intervention. A method must be sought to adjust the repair according to the extent and direction of the crack as it was revealed during the intervention.

The resin solution was therefore rejected. The fault in the column lay in an isolated primary supporting member, the collapse of which would be disastrous. The 'glueing' idea did not give any guarantee of success in this particular case.

Many permutations of strapping or tying were also considered. There are many instances of historic structures being repaired using external, exposed, metal straps. This method generally fulfils a need in preventing collapse, and is usually the least expensive solution, but it lends nothing to considerations of aesthetics. Sometimes a neater solution can be found.

Remembering that the crack was at 45 degrees with a propensity for the uppermost part of the column to slide down the crack surface, straps or tie rods could resist the horizontal component of this force, leaving its vertical component to safely continue down the column. This was a simple idea, but the horizontal forces in any tie rods have to be anchored into the stone. Ideally, any straps have to be hidden. If the halves of the column were to be strapped together the steel strapping would have to be hidden within deep grooves carved around the fluted stone column and disguised, then covered with a plastic (lime mortar) repair made to look like the original fabric.

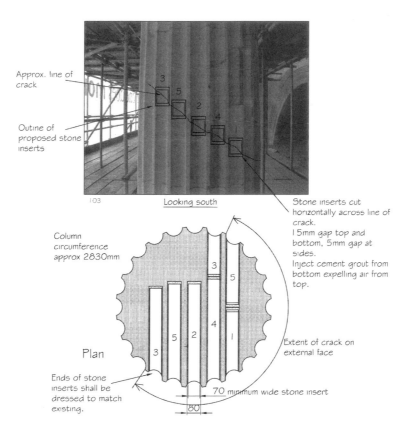

Figure 4. Plan and elevation of proposed works (photograph by Les Ayling, English Heritage, drawing copyright © English Heritage).

Figure 5. Close-up view into the pocket, showing the line of the crack (photograph by Les Ayling, English Heritage).

In order to achieve this, an unacceptable amount of the viable original stone would have to be cut away to provide enough room to bury not only the strap but also its fixings.

If the sections of stone on each side of the crack were to be tied together with tie rods then the patras plates would also have to be hidden. These would have to be sunk into pockets in the column, deep enough and large enough to accommodate the plates and bolt heads.

As referred to previously, the crack line resulted in a circular, tapered wedge-shaped piece of stone above the crack. Horizontal tie rods in pockets and straps in grooves could not be anchored into such a small, tapering piece of stone alone. This was considered to be unacceptable.

Therefore, all strapping or tying proposals proved inadequate due to the size and position of the anchorage plates that would be required and the amount of stone to be cut away in order to recess and disguise the intervention.

The chosen solution

The inspiration for the chosen solution developed from a 1993 advertisement showing a 'new concrete cutting tool'. A chainsaw was depicted cutting neat rectangular holes through a reinforced concrete wall (Fig 3). If this tool could cut through concrete, what problem would Portland stone hold? The solution, borne out of the apparent ease of cutting small, deep, rectangular pockets, was to cut out small fingers of stone containing the sloping crack and replace them with new stone inserts (Fig 4). A pocket included the crack on both of its sides and by these means the extent of the crack could be inspected (Fig 5). The new inserts would transfer the load from above to below the crack through a vertical load path, changing the sideways component of the force back into a vertical one. The inserts would extend at least to the centre of the column to meet, where necessary, the ones cut from the other side. Using this approach the variations in the line of the crack across the column could be accommodated. Design checks showed that the entire load could be carried on the combined strength of these inserts and the remaining uncracked stone, ignoring the small sections of the crack left between. The width of the inserts was chosen so that they would fit within the carved flute in the column and thus retaining the ribs, an important feature of the column. Their height matched the size of the chainsaw, approximately 108 mm (4 ¼ in).

WORK ON SITE

WT Specialist Contracts, who had the appropriate equipment, were chosen to carry out the cutting, drilling and

Figure 6. The ICS® chainsaw in action, cutting a pocket (photograph by Les Ayling, English Heritage).

grouting of the stone inserts under the control of the stonework subcontractor, Stonewest Ltd. A temporary works scheme to provide structural support for the roof, entablature and column capital was supplied by Mansell PLC, the main contractor, and threaded through the existing access scaffold. To ensure the safety of the structure and to minimize disturbance to the crack line, the pockets were cut in a sequence similar to that used when underpinning walls. A numbered sequence prevented two adjacent pockets being cut in succeeding visits. The cutting and grouting of each insert was then separated by a five-day curing period, allowing the new load-bearing element to become established and to ensure that its grouting was not damaged by subsequent cutting.

The chainsaw was a lightweight, hand-held, hydraulically-powered cutting unit and in this instance guided by the skill of the operator using a simple timber guide fixed to the scaffolding (Fig 6). Initially, a 100 mm-deep (4 in) pocket was cut to check the alignment of the crack. If the crack had turned out to be shallower than expected or if it did not follow a direct line across the column then, using this machine, adjustment could easily be made to the height of the pockets. This meant the height of the pockets could be increased once cutting had commenced in order to follow the crack. The saw cut across the cracks leaving a neat vertically sided pocket in which, with the aid of a low-pressure water supply to flush out the debris, the crack lines could be clearly seen (Fig 7). The saw had a length of approximately 450 mm (18 in) and was suitable for most of the pockets. Where a deeper length was required the pocket was extended by core drilling the end of the sawn hole and trimming it with a chisel specially shaped for this purpose, an operation that turned out to be simpler than envisaged. The grout was carefully chosen to give an early strength, minimizing disruption to the existing contract, to have minimal shrinkage and good flow characteristics. The mix was chosen because of the structural nature of its application, and, for logistical reasons, diverted from English Heritage's general tenet of deploying relatively weak, flexible, lime-based mortars and grouts. The object here was to completely surround the new stone inserts with a grout that could, within a reasonable time, safely support load with minimal shrinkage while the cutting sequence was carried out, to restore the column's structural integrity. The grout consisted of a high-strength fast-setting blend of ordinary Portland cement to BS12, pulverized fuel ash to BS 3892 and plasticized controlled expansion additives. The PFA, as a pozzolan, ties up all free lime in the cement binder and enhances the speed and efficiency of the setting process.

Because the 'joints' around each insert were designed to be such a tight fit, the grout's free-flowing characteristics enabled it to reach all interstices effectively without leaving too much to be pointed up to match the existing stone on completion. The grout was pumped through injection tubes to the back of the inserts, thus flowing forward expelling air as it travelled (Figs 8 and 9). The inserts were finally faced off to match the column flutes and pointed to match the surrounding stone.

Figure 7. Inserts ready for facing off, a workman inspecting the crack line with a powerful torch (photograph by Les Ayling, English Heritage).

Figure 8. Close-up of the stone insert (photograph by Les Ayling, English Heritage). See Colour Plate 37.

Figure 9. One to go! The north face after grouting four inserts (photograph by Les Ayling, English Heritage).

Figure 10. The completed repair (photograph by Les Ayling, English Heritage). See Colour Plate 38.

CONCLUSION

The fine crack line, which was discovered only after close inspection and part-way into the conservation and restoration contract, was repaired using a technique sympathetic to conservation ideals. It achieved the objective of minimum disturbance leaving minimal evidence of the intervention and gave the least disruption to the existing contract works. The difficulty of handling large blocks of stone was resolved by the use of small stone inserts that could be picked up in one hand. The chosen solution was the culmination of a systematic appraisal of possible options and the critical assessment of design and construction aspects. Following the completion of the conservation works, the Wellington Arch will be removed from English Heritage's 2001 *Register of Buildings at Risk* and with its new floodlighting will again stand out as one of London's most well-known landmarks (Fig 10).

ENDNOTES

1. It was resited near the garrison church at Aldershot, Hampshire.
2. Information from Stephen Brindle. See also Brindle 1999, and Brindle 2001.
3. The metal detectors used were a reinforcement cover meter type, which is restricted to ferrous metals only, and a long arm circular search head type, which can discriminate between other metals. See equipment list below.

BIBLIOGRAPHY

Brindle S, 1999 *The Wellington Arch, Hyde Park Corner, A History*, unpublished internal report, London, English Heritage.

Brindle S, 2001 The Wellington Arch and the western entrance to London, *The Georgian Group Journal*, **xi**.

British Standards Institution, 1992 *BS5628: Part 1 Use of Masonry, Part 1 Structural Use of Unreinforced Masonry*, London, British Standards Institution.

The Institution of Structural Engineers, 1996 *Appraisal of Existing Structures*, second edition, London, The Institution of Structural Engineers.

Warland E G, 1929 *Modern Practical Masonry*, London, The Library Press Ltd.

ADDRESSES

Main contractor: Mansell PLC, Roman House, 263–269 City Road, London, EC1V 1JX, UK; Tel: + 44 20 7490 1220; +44 207 490 0079.

Stone cleaning, repair and replacement: Stonewest Ltd, Lamberts Place, St James's Road, Croydon, CR9 2HX, UK: Tel: + 44 20 8684 6646; Fax: +44 208 684 9323.

Chain saw cutting, drilling and grouting: WT Specialist Contracts; Tel: + 44 1273 479764; Fax: +44 1273 479765, www.wtgroup.co.uk

EQUIPMENT

Chain saw: ICS® Blount Europe, Nivelles, Belgium; Tel: + 32 67 887628: UK contact: + 44 7977 282518, www.icsbestway.com

Metal detectors: Micro Covermeter, from Kolectric Ltd; Tel: + 44 1844 261626; Fax: + 44 1844 261600.

cs1220xdp, from C-Scope International Ltd, Kingsnorth Technical Park, Wotton Road, Ashford TN23 6LN, UK; Tel: + 44 1233 629181; Fax: + 44 1233 645897.

AUTHOR BIOGRAPHY

Les Ayling is a conservation engineer from English Heritage's Conservation Engineering team who was the project structural engineer for the conservation project at the Wellington Arch. He provides, as part of English Heritage's team of engineers, structural and civil engineering conservation advice in protecting the historic built environment.